ENVIRONMENTAL LAW AND AMERICAN BUSINESS

Dilemmas of Compliance

ENVIRONMENT, DEVELOPMENT, AND PUBLIC POLICY
A series of volumes under the general editorship of
Lawrence Susskind, *Massachusetts Institute of Technology*

ENVIRONMENTAL POLICY AND PLANNING
Series Editor:
Lawrence Susskind, *Massachusetts Institute of Technology, Cambridge, Massachusetts*

BEYOND THE NEIGHBORHOOD UNIT
Residential Environments and Public Policy
Tribid Banerjee and William C. Baer

CAN REGULATION WORK?
Paul A. Sabatier and Daniel A. Mazmanian

ENVIRONMENTAL DISPUTE RESOLUTION
Lawrence S. Bacow and Michael Wheeler

ENVIRONMENTAL LAW AND AMERICAN BUSINESS
Dilemmas of Compliance
Joseph F. DiMento

THE LAND USE POLICY DEBATE IN THE UNITED STATES
Edited by Judith I. de Neufville

PATERNALISM, CONFLICT, AND COPRODUCTION
Learning from Citizen Action and Citizen Participation in
Western Europe
Lawrence Susskind and Michael Elliott

RESOLVING DEVELOPMENT DISPUTES THROUGH
NEGOTIATIONS
Timothy J. Sullivan

Other subseries:

CITIES AND DEVELOPMENT
Series Editor:
Lloyd Rodwin, *Massachusetts Institute of Technology, Cambridge, Massachusetts*

PUBLIC POLICY AND SOCIAL SERVICES
Series Editor:
Gary Marx, *Massachusetts Institute of Technology, Cambridge, Massachusetts*

ENVIRONMENTAL LAW AND AMERICAN BUSINESS

Dilemmas of Compliance

Joseph F. DiMento
University of California, Irvine
Irvine, California

Plenum Press • New York and London

Library of Congress Cataloging in Publication Data

DiMento, Joseph F.
  Environmental law and American business.

  (Environment, development, and public law. Environmental policy and planning)
  Bibliography: p.
  Includes index.
    1. Environmental law—United States. 2. Pollution—Law and legislation—United
  States. I. Title. II. Series.
  KF3775.D5 1986   344.73'046   86-75
  ISBN 0-306-42168-2   347.30446

© 1986 Plenum Press, New York
A Division of Plenum Publishing Corporation
233 Spring Street, New York, N.Y. 10013

Printed in the United States of America

To little man, J. L. D., with love

PREFACE

We are in the second decade of modern environmental law. By some indicators this body of regulation has matured greatly. We can point to statutes and codes at the federal, state, and local levels which address almost every conceivable form of pollution and environmental insult. Yet, despite the existence of this large body of law, despite considerable expenditures on enforcement, and despite the energetic efforts of people sympathetic to environmental objectives, violations are numerous. Serious pollution problems are commonplace. Love Canal, the Valley of the Drums, Times Beach, and Stringfellow Acid Pits epitomize the national environmental quality challenge. Daily, a major illegal disposal of hazardous waste is recorded; a new mismanaged dump site is discovered; a toxic substance is found in our drinking water; or a failure to meet a water or air quality standard is identified.

Many of these violations involve American business. Failures to comply are of several types. A small businessman in Pennsylvania mistakenly allows a spillover of a pollutant into a protected stream. An industrialist in the Midwest adds to his fortune by illegally dumping dangerous chemicals. A series of errors by several firms, some of which no longer exist, combine to create a health-threatening conflagration on the West Coast. An automobile company interprets one of the almost innumerable air pollution rules differently from government: It produces a car which the government says fails to comply with the Clean Air Act.

Part of the challenge of achieving compliance derives from this diversity of business actions and part from the different interpretations of compliance by corporate executives, governmental lawyers, environmental advocates, and private citizens. So, too, assessments vary on the wisdom of promoting strict adherence to environmental rules over other national goals, including economic growth and employment.

Environmental law is at a critical juncture in its history. We clearly must reduce the incidence of violations. But we also must respond to legitimate economic and other interests, including that of efficiency in government.

This book offers policy reforms that aim to achieve the goal of inducing business to comply with reasonable environmental law. By describing classes of illegal acts, linking them to characteristics of noncomplying companies and to types of rule development, and critically reviewing economic and legal strategies to influence the firm, I hope to add to our knowledge of deterrence of rule violations and environmental crime.

Environmental law is a window on the regulatory process. My second aim is to cut into the debate about the wisdom, efficacy, and effects of health and safety rules in general. Attention to enforcement, to communication of law, and to characteristics and motivations of the groups that make, defy, and comply with regulation uncovers implications for many areas of regulation.

What function can the criminal law play in promoting compliance? Why do business responses to regulation differ? Are there relatively inexpensive ways to promote regulations while respecting rights of privacy and property? Do rules that are seen as rational, fair, and cost-effective invite compliance? I address these questions and attempt to identify traits of government associated with legitimate and authoritative law and types of enforcement which make legal action work.

There is no one segment of the literature or one classic case that identifies why organizations comply. No one respondent group has a persuasive insight into compliance. But several bodies of writing do describe the firm's response to law. Also, business executives, government attorneys, agency officials, and private citizens—the respondents in our study—offer perspectives on what makes environmental law effective. And several cases, such as the four that begin this book, demonstrate how businesspeople react to regulatory law. This volume calls on these writings and experiences to suggest means of realizing reasonable compliance with reasonable law.

JOSEPH F. DiMENTO

ACKNOWLEDGMENTS

This book has benefited from the assistance of several people. In developing the framework for compliance and addressing the approaches to data collection, I gained important insights from Gilbert Geis. He also assisted in the interviews in the automotive industry and made extensive valuable comments on drafts of the manuscript, as did John Monahan. James Krier asked pointed questions about my focus on compliance and on its significance; these made the project more objective. Thomas Anton identified an important gap in the original compliance framework. Steven Colome provided useful perspectives on the environmental sciences as they influence the regulatory process. I, of course, remain fully responsible for any deficiencies in the framework.

Wesley Marx helped me with suggestions on the generalizability of my findings and on writing style. I was assisted in bibliographical work and data collection by William Lambert, Terri Port, and Dean Hestermann, graduate students in the Program in Social Ecology at the University of California, Irvine. Pat Ward, Beverly McKinney, and Sandra Wickland provided invaluable production assistance.

David Swan, Kennecott Copper; Herbert Misch, Ford Motor Company; Richard Davis and David Potter, General Motors; and Christopher Kennedy, Chrysler Corporation stand out for their helpfulness in arranging interviews in their companies. Also, I thank all those persons who allowed me to investigate their attitudes toward environmental law and their analyses of the cases in which they were involved.

The *Louisville Courier Journal* and *The News and Observer* and *The Raleigh Times* (Raleigh, North Carolina) provided valuable background information on midnight dumping cases.

I received financial assistance from the University of California, Irvine. Timely grants on two occasions enabled me to complete the data collection. I wish to thank the Kellogg Foundation for funding under the National Fellowship Program that allowed me to interview government and industry officials and do other work on the scientific bases of environmental regulation.

CONTENTS

LIST OF ABBREVIATIONS

CEO	Chief Executive Officer
DOI	United States Department of the Interior
DOJ	United States Department of Justice
EDF	Environmental Defense Fund
EPA	United States Environmental Protection Agency
MEPA	Michigan Environmental Protection Act
NAAQS	National Ambient Air Quality Standards
NRC	Nuclear Regulatory Commission
NRDC	Natural Resources Defense Council
NSPS	New Source Performance Standards Under the Clean Air Act
OMB	Office of Management and Budget
OSHA	Occupational Safety and Health Administration
RCRA	Resource Conservation and Recovery Act
SCAQMD	South Coast Air Quality Management District
SIP	State Implementation Plan
Superfund	Comprehensive Environmental Response, Compensation, and Liability Act Provision
TCE	Trichloroethylene
TSCA	Toxic Substances Control Act

CHAPTER 1

MUSHROOMS AND MERCURIES

Cases of Noncompliance

Case 1: A Mushroom Farmer Goes to Jail for a Water Runoff from a Holding Tank

James and Guido Frezzo are mushroom farmers in Avondale, Pennsylvania. They inherited this business from their father, who began it in the 1930s after he decided to give up quarrying in western Pennsylvania. Since then all family members have been mushroom farmers or have been married to mushroom farmers.

The hilly, rural region in which the Frezzo brothers farm is in transition. Commuting executives are eager to make this attractive area their home. The district is touched in many ways by the mushroom industry; roadside stands and restaurants that cater to tourists promote tarts and quiches and other mushroom products.

Mushroom farming, however, is not a pretty business. One principal activity involves making the compost in which mushrooms are grown in dark shacks. The compost contains hay, horse manure, chicken manure, a kind of chocolate, and gypsum. These materials are mixed into windrows called ricks which must be watered and turned periodically so that bacteria will develop properly. Tradition dictates the processes which are used. The scene is one of mud and manure.

Water runoff from the composting operations was the subject of the litigation involving the Frezzos. The runoff was designed to be recirculated into the compost through a 114,000 gallon concrete holding tank built in the 1950s. But on a number of occasions in the summer of 1977 and the spring of 1978, a Chester County Health Department investigator, Richard Casson, observed that the White Clay Creek near the Frezzo property was being polluted; he collected samples of runoff from a channel box that carried water from the Frezzo's land through a neighboring farm. Casson also entered the property with a search warrant and with witnesses. The government made its charges with this information. The specific allegation was that these discharges occurred without a permit required under the Federal Water Pollution Control Act Amendments of 1972.[1] The Frezzos had never applied for a permit.

The United States argued successfully that permits were needed even when no effluent standards had been made applicable to a pollution source. It relied on the legislative history of the Water Act. The case was criminal in nature, and the government did not first bring a civil action. It argued that no such action, nor any official administrative proceeding, was necessary. A criminal case could be brought by the EPA administrator from the very start.

For evidence the government submitted samples of the discharge from the Frezzo property, photographs showing how runoff flowed into the affected waterways, expert testimony on hydrology, and testimony that the company had been visited earlier by state environmental officials, this last to prove that the brothers knew that the holding tanks were inadequate to contain all runoff.

Guido Frezzo would not talk with me about this case. Convinced that the government "is out to get" him, he has lost faith in all levels of government. Guido's brother, James, describes the background to the prosecution this way:

> The day they came in there was a deluge, a downpour. . . . My tank was overflowing. Of course I was pretty upset . . . . The material that we inadvertently made go into the stream is a biodegradable material that wouldn't hurt a soul. It's not something like you'll see in these industrial plants or a highly poisonous type of material.[2]

To James Frezzo the approach the government used to make him comply was irrational and unfair.

> One of our biggest problems was that no state or county agency would give us a definite guideline as to how to control our problem. . . . If we'd known before time, it would be ridiculous not to come up to the provisions of the state and federal government.[3]

Government prosecutors did not see the problem as so complex; this was a case of willful discharge by a compost producer who clearly had the means to install proper control:

> There is a problem because of all that manure. The simplest way to avoid it is to put some kind of roof over the compost, so the rainfall doesn't cause a problem. . . . They did put in additional holding tanks with larger capacity, but this is not enough.[4]

---

[1] 33 U.S.C. §§ 1311(a), 1311(c), (1970 ed. Supp. IV).

[2] Interview with James Frezzo, Dec. 28, 1981.

[3] *Id.*

[4] Interview with Bruce Chasan, Jan. 8, 1985.

Frezzo said that prior to his indictment he had not even known the Federal Water Pollution Control Act existed. The *Frezzo* motion to discuss the indictment contended that the Frezzos "cannot have violated the Act because the EPA has not yet promulgated limitations which they can be held to have violated."[5] Government inspectors had entered the Frezzo property. "Over the years, they'd stop in and would see the type of controls which we were using, but they wouldn't say they were . . . sufficient or insufficient."[6] According to Frezzo, the government would not recommend an engineering firm to help him; it was interested only in prosecuting. Nonetheless, testimony accepted by the federal district court concluded that

> as far back as 1970, the defendants in this case had been investigated, visited and confronted by a number of state and county employees concerning the fact that the stream in question was being polluted by runoff from the compost operation. . . . A letter dated October 4, 1971 from James to the DER [the Pennsylvania Department of Environmental Resources], introduced as an exhibit by the Government, indicated that James knew that the holding tank was inadequate.[7]

In fact, in a letter to the Pennsylvania Department of Environmental Resources in 1971, James Frezzo wrote:

> I planned on an additional tank for water storage. This would help me better contain run-off and rainwater. . . . I am still considering the installation of another holding tank.[8]

Frezzo believes that the choice of targets for prosecution had been based, at least in part, on the fact that the mushroom industry had just experienced a very profitable period. He added that "DuPonters," the city executives who were moving to Avondale, disliked mushroom agriculture. The government seemed less concerned with compliance than with prosecuting. "They were just sitting back waiting to gain information. . . . All along, it didn't seem like they were interested in cleaning up the environment . . . in telling me how to control my problem. They were more interested in convicting James and Guido."[9]

The government did not focus only on the Frezzos; four other indictments brought at the same time involved pollution incidents from compost manufacturing.[10] But the government rationale was not based on excess profits of the industry; rather, the compost industry was seen as "notorious" in its disregard for pollution control.[11]

---

[5] *United States v. Frezzo Brothers*, 602 F.2d 1123, 1127 (1979).

[6] Interview with James Frezzo, Dec. 28, 1981.

[7] *United States v. Frezzo Brothers, Inc.*, 461 F. Supp. 266 (1978) at 270.

[8] 461 F. Supp. 266 (1978).

[9] Interview with James Frezzo, Dec. 28, 1981.

[10] Interview with Bruce Chasan, Jan. 8, 1985.

[11] Interview with Bruce Chasan, Jan. 8, 1985.

Although admitting that runoffs did occur at times, Frezzo emphasized that "most of the time" the operations were in compliance. The government read this differently. It interpreted the statement of control "95% of the time" as an admission that the Frezzos were not completely controlling their run-off problem:[12] "What they meant was, when it wasn't raining they were in compliance."[13]

The district court found that on "each of the six dates charged in the indictment, runoff from the compost pile made its way into the storm water runoff system and was carried through the pipe into the tributary of the White Clay Creek."[14] And the court also found that the chemical and bacteriological tests were properly conducted and did establish that the discharges were "sewage." As to the source of the runoff:

> One witness testified that he had actually walked along the path of the pipe from Penn Green Road to the pipe's end at the tributary and that there were no other mushroom manufacturers upgrade of the channel box. He further testified that results of analyses of samples from the channel box and from the White Clay Creek substantiated his conclusion that the pollution in the White Clay Creek came from the Frezzo property.[15]

James Frezzo believes that government acts that educate and that do not treat violating business activities as crimes can best serve compliance.

> If the environmental control agencies would come forward and tell business this is what we want, this is what we need, and then put a deadline on it . . . you'd have to be a complete fool not to comply. You cannot operate a business in underhanded methods daily . . . . So rather than spending all the time and money trying to put a case together like you're going first to prosecute us, if they would have loaned that amount of money when I asked for it, we would have solved the problem then.

> The Environmental Protection Agency should have had an area meeting with the mushroom industry, explained to the growers: "This is what the problem is, this is what the guidelines are" . . . give the grower 90 or 180 days to come in compliance. I compare it to driving down the highway. . . . You see the sign that says 55 miles an hour, and you're driving 65, you know you're breaking the law. But if there's no sign, how do you know when you break the law?[16]

Although the appropriateness of criminal sanctions in pollution cases is not a matter of consensus among government officials, one of the government prosecutors in the Frezzo case was quite clear on the matter:

> This is my opinion. I think that a lot depends on the prosecutor's inclination. I developed a view that pollution incidents were almost always willful. . . . Some-

---

[12] *United States v. Frezzo Brothers, Inc.*, 461 F. Supp. 266 (1978), n. 4.
[13] Letter from Bruce J. Chasan to Joseph F. DiMento, July 9, 1985.
[14] 461 F. Supp. 269 (1978).
[15] *Id.*
[16] Interview with James Frezzo, Dec. 28, 1981.

where the operator has made a decision not to put money in pollution abatement. The parties that don't spend money are exercising a willfull form of tax evasion. In a case like Frezzo, they all knew what their problems were, and all believed that enforcement would never be used. And they thought they could forget about compliance.[17]

The Frezzos declined to plea bargain.[18] The government successfully made its case before a jury, the Frezzo brothers' company was fined $50,000 and the brothers were individually fined and sentenced to imprisonment for thirty days. The brothers attempted to characterize the discharge as an agricultural activity and therefore exempt from EPA permit requirements. But the court concluded the enterprise was manufacturing in nature and that the Frezzos had failed to establish that they had, in fact, relied on their understanding of that exemption. On appeal, the Frezzos' conviction was affirmed.[19] A case that went twice to the United States Supreme Court and three times to the court of appeals was finally resolved.

Case 2: A Businessman Is Convicted of Disposing of Toxic Chemicals into the Louisville Sewer System

On October 5, 1981, without comment, the United States Supreme Court refused to hear an appeal of the first criminal conviction by a jury for a violation of the Federal Water Pollution Control Act. Allowed to stand was a record sentence in a pollution case: two years in prison and a fine of $50,000 for Donald E. Distler, an officer of Kentucky Liquid Recycling, Inc., of Louisville, Kentucky.

In March, 1977, a pungent-smelling substance was discovered in the Morris-Foreman Wastewater Treatment Plant in the metropolitan sewer district of Louisville. With a dozen and a half workers suffering from headaches, nausea, and loss of memory, the district closed the plant. A hundred million gallons of raw sewage poured into the Ohio River every day.

Analyses by the EPA concluded that the odors were caused by two highly toxic chemicals: hexachlorocyclopentadiene (hexa) and octachlorocyclopentene (octa).

Who were the illegal dumpers? No suspects were immediately identified. The Federal Bureau of Investigation became involved in April, 1977. Subsequently, the investigation involved the United States Attorney General, the Kentucky Division of Water Quality, the EPA, the Kentucky State Police, and city and county police as well as city firemen and the army's 194th Armored Brigade

---

[17] Interview with Bruce Chasan, Jan. 8, 1985.

[18] Interview with Bruce Chasan, Jan. 8, 1985. The plea bargain would have involved a fine only.

[19] *United States v. Frezzo Brothers,* No. 82-1494, 703 F. 2d 62 (3rd Cir. Mar. 29, 1983), 13 ELR 20584; and No. 82-2136, 13 ELR 10382 (Oct. 4, 1983).

from Fort Knox. The FBI alleged that octa and hexa were dumped in city sewers sometime around March 1. The bureau narrowed the source to an area bounded by Twenty-sixth and Twenty-eighth streets, West Broadway and Howard streets, four city blocks which contain an old distillery.

Circumstantial evidence led to the initial charge: Distler owned storage sites in the area; at least one contained hexa and octa. The hexa stored in one Distler site was produced by Velsicol Company of Memphis. Velsicol allegedly shipped the hexa to the site, where it was improperly stored for a period of time. Distler had not completed an application to incinerate at least one of the chemicals, but in August, 1976, he had contacted the Kentucky Department of Natural Resources "to determine what permits were required in order to operate a liquid waste disposal site,"[20] and he had hired an incinerator designer.

The federal grand jury indictment that followed on the FBI charges included a conspiracy count and alleged that, on a number of occasions between September and March, the named defendants had illegally dumped pollutants into the city sewers. The firm's president, Donald Distler, and the secretary-treasurer and one other employee were charged with violating federal water pollution control laws[21] of which Distler claimed complete ignorance.[22] The defendants other than Distler were ultimately either dropped from the case or were acquitted. Indictments linked Distler's polluting activity to the problem in the Morris-Foreman Wastewater Treatment Plant. A separate count charged Distler with illegal discharges of dangerous substances in a field his parents owned in Jefferson County, Kentucky. As later became known, the defendant stored almost 2,000 barrels of reusable chemicals, solid waste, and unusable liquid waste in various area sites.

The defendant pleaded not guilty. The trial began on November 9, 1978, in the United States District Court in Louisville. A month and a half later, Distler was convicted on two pollution counts and acquitted on the conspiracy count.

One juror stated that the jury was first divided over Distler's guilt.

> The jury was split with 6 votes to convict and 6 votes to acquit Friday evening. It narrowed down to only 2 favoring his acquittal on Friday and then this morning we found him guilty. There was no direct evidence to show that he was guilty. It was all circumstantial. But in the end we felt there was enough evidence to implicate him.[23]

Nobody had observed the alleged dumping, although witnesses saw a person who looked like Distler in the area during the pollution period "washing . . . tankers out, and on some occasions a hose was seen extending from the tankers into the sewer."[24] But this testimony alone might not have been ade-

[20] *United States v. Distler,* 671 F.2d 954 (1981), at 756.
[21] Specifically, two violations of the Federal Water Pollution Control Act, 33 U.S.C. § 1251 *et seq.*
[22] Interview with Donald E. Distler, Apr. 22, 1985.
[23] Quoted in *The Louisville Courier Journal,* Dec. 27, 1978, at E7.
[24] *United States v. Distler,* 671 F.2d 954 (1981), at 963.

quate to lead to conviction. United States Attorney David Everett relied on evidence obtained through sophisticated chemical analyses, "scientific fingerprinting" including gas chromatography tests and other chemical identification processes. Chemical profiles were used to match samples of oil mixed with "PCL bottoms" taken from the entry chamber of the sewer plant with samples taken from Distler's Liquid Recycling Company.

The analyses can detect chemicals in quantities as small as one billionth of a gram. The scientific data are formated in a printout that looks something like an electrocardiogram. Pollutant samples that cannot be seen by the human eye become visible and incriminating when their profiles are produced. The test relies on the characteristic boiling points of substances: the temperature at which a sample substance evaporates as well as the quantities given off can be used to identify its components. Since oil samples are unique and refined oil retains its unique character, oil sample matching was used for comparing the sewer samples with the Distler samples.

Distler argued both in the court and after his release from prison that the evidence was flawed and insufficient. He challenged the expert testimony in the case arguing that the use of gas chromatography was not a generally accepted method of matching samples and specifically that it is not a technique transferable to analyzing sewer samples. The courts rejected these arguments.[25]

On September 14, 1979, Donald Distler was sentenced to two years' confinement and a $50,000 fine. U.S. District Judge Charles Allen said:

> No defendant to come before this court has exhibited a more callous and flagrant disregard for the safety and lives of vast numbers of citizens of this area.[26]

The judge's comment reflected the U.S. Attorney's written statement (characterized as "unusual" by the press):

> The acts for which this defendant stands convicted constitute what the United States considered to be one of the most egregious white-collar crimes.[27]

Judge Allen also stated: "Businessmen and industries who pollute our environment are guilty of great crimes against man, nature, and themselves." If allowed to continue, these crimes would create "effects . . . irreversible by any known technology."[28]

Despite the unprecedented nature of the case, not all observers thought the sentence sufficient. Robert W. Keats, attorney for the Metropolitan Sewer District, felt, in view of the damages from the dumping, that the sentence was too light. The district incurred considerable costs; its employees were subjected to

---

[25] *United States v. Distler,* 671 F.2d 954 (1981).
[26] Quoted in *The Louisville Courier Journal,* Sept. 15, 1979, at 7.
[27] *Id.*
[28] *Id.*

very dangerous, difficult, and unpleasant cleanup activity. Workers had to enter the sewers on summer days at temperatures that reached 115 degrees Fahrenheit. Dressed in futuristic, astronaut-type protective gear, the crews shoveled the chemical residue into pails and barrels to remove it from the sewers. The impact of daily discharges into the Ohio River of 100,000,000 gallons of untreated sewage were never fully established.

Keats may have had visions of sick employees in the district, his work force standing in three feet of water as they removed the unpleasant goo; and he may have been thinking of his own increased caseload resulting from suits brought by downstream municipalities. The city of Evansville, Indiana, for example, argued that it should be compensated for damages to its own water systems, for excess costs and employee travel required to undertake testing related to the Louisville runoffs.

There were other side effects of the illegal disposal. Several Louisville companies purportedly took advantage of the plant's closing to dump untreated oils, greases, and solvents into the sewers.

The *Louisville Times* looked into the costs of the *Distler* case.[29] There was a range of estimates, with $1.5 million the figure generally accepted. This price tag did not include expenses incurred by the EPA for laboratory and other tests that accounting methods do not allocate to individual projects. Related to the *Distler* case were the salaries and wages of FBI agents who worked on the case ($51,000); medical tests ($85,000); the costs of U.S. marshalls (almost $2,000); the disposal of the wastes ($25,000); expenses of jurors (over $10,000) and the considerable costs of sewer repair.

Distler throughout maintained his innocence.[30] He said: "Although we have exhausted our appeal remedies, we maintain to this hour that we are free of any wrongdoing of the charge, even if viewed in the light most favorable to the Government."[31] Distler concluded that the government "was just out to get a conviction."[32] There was no reason for him to illegally dispose of the chemicals, he has argued, since at the time of the alleged dumping only ten days remained until he had the capacity to begin legal incineration of the materials. Distler also claims that he had additional storage space and that the only proper disposal method is incineration.

Distler concluded that he was improperly treated in part because he was so "hardheaded" throughout the proceedings. He refused to plea bargain[33] and he did not make friends in the courtroom. He thought "there was no way in the

[29] *Louisville Times*, Mar. 6, 1979, at E4.
[30] Letter of Donald E. Distler to Joseph F. DiMento, Sept. 22, 1982, and interviews with Donald E. Distler, Apr. 19 and Apr. 22, 1985.
[31] Letter of Donald E. Distler to Joseph F. DiMento, Sept. 22, 1982.
[32] Interview with Donald E. Distler, Apr. 22, 1985.
[33] *Id.*

world that he would be convicted,'' but he underestimated the resources that the government would allocate to insure a conviction:

> The government has all the money in the world. Witnesses will say what the government wants them to say. You've got all types of enforcement agencies: These guys are gods (to the witnesses).[34]

Distler also believed that the jury was totally incapable of understanding the scientific evidence required to establish noncompliance: "These were poor people . . . who slept during the trial: it was way over their heads."[35]

Case 3: A Hazardous Waste Site Becomes An Inferno and No One Appears Accountable

On May 4, 1981, the State of California filed a civil suit against the operator, current owner, former owner, and former operators of a hazardous waste facility in Santa Fe Springs, an industrial city in southern California. The suit represented a milestone in a ten-year history of attempts by public agencies to compel cleanup and proper disposal of 14,000 drums of waste stockpiled at the site. EPA officials compared the storage site (locally referred to as the Inmont site) in size and potential hazard to that of the Chemical Control Corporation in Elizabeth, New Jersey, the 1979 scene of the nation's largest waste site fire.

Just two months after the filing of the suit, an arsonist's torch started an inferno at the Inmont site which lasted for several days.

The site, a concern of the city for over a decade, was operated in the late 1960s by General Disposal Company. At first the city wanted the site fenced and properly maintained. Enforcement efforts were based on planning, zoning, fire and building codes related to storage of barrels, fencing, and adequacy of parkway. Under this approach the cycle of notification and perfunctory compliance was regular.

Over the years, as barrels accumulated, the city became concerned about fire hazards. Contents of the barrels were unknown, although explosives were suspected. In 1975 the city filed suit to force clean up. The current operator promised—but did not carry out—compliance. The city filed suit again to gain temporary and partial compliance.

A city official described the response of one principal in the case:

> Mr. Stankovich was slapped with a $400 fine for the violation of the fire code. It virtually didn't have any effect at all. He didn't even blink. . . . We began to realize that, under our fire . . . zoning and planning codes . . . we only had misdemeanor enforcement powers and did not have powers other than the power of persuasion and

---

[34] *Id.*

[35] *Id.*

the power of fear and going to court and had very minimal authority to force the issue.[36]

The city decided to enact a property maintenance ordinance with provisions for notice and hearing to force a property owner to abate a public nuisance. The city could clean up the site and impose a lien if the polluter did not comply.

A buyer for the site was found in April, 1979. The new owner, William Boyer, was reportedly assured by the seller of the land (from whom Stankovich, the operator, had leased the property) that the previous operator would clean up the property before ownership passed.

When the new owner took possession, 14,000 barrels of unidentified materials were stored on the site. The few that were labeled had vague descriptions such as "slop" and "dirty wash." The barrels, stacked four feet high, were located near a flood control channel. Many were corroded and bloated. Some lay on their sides, leaking their contents into the channel. Protecting the surrounding area was a six-foot chain link fence in poor repair. A caveat, "Danger—Hazardous Materials," was posted at widely spaced intervals on the fence.

The city directed Boyer to clean up the site. Boyer estimated that he could do so for around $30,000. Concurrently, the State of California Health Department cited Stankovich for illegally dumping toxic wastes from a pump truck along the roadway.

Neighboring property owners began to recognize other impacts. For instance, banks were reluctant to make property loans in the area until they were assured that the drums would be removed.

The case became more complicated when Santa Fe Springs tried to expedite its cleanup order. Stankovich successfully sought to enjoin both the city and Boyer from removing any barrels, arguing that their contents were valuable. Meanwhile, materials suspiciously familiar to those found on the site continued to be dumped along area roadsides. Santa Fe Springs officials said that there was a "rather suspicious correlation" between these dumpings and the required cleanup of the site in question.

Stankovich then sued the city. He charged an illicit compromise involving Boyer and the city and accused both of ignorance of proper disposal methods. In a countersuit, Boyer claimed that Stankovich was creating a nuisance, diminishing the value of his land, and subjecting him to possible third-party lawsuits. Boyer then sued the original owner, an estate. At one point, there were five lawsuits totalling over $200 million in alleged damages.

The site remained poorly managed and the city considered cleaning it; but the initial cost estimate, $30,000, was too low. The city sought consulting assistance. According to the assistant city manager: "Everyone in the world who

---

[36] Interview with Fred Latham, Assistant City Manager, Santa Fe Springs, Apr. 21, 1982.

thought about cleaning up a site gave us a proposal. The proposed cost ranged from $200,000 to 2,000,000 and involved various types of 'star wars' equipment.''[37] Ideas ranged from highly technical systems for cleanup to one in which a single operator promised to come in with a vacuum truck and clean out the barrels quickly. One idea—selling the wastes abroad—became known as "The Taiwan Deal."

A consulting engineering company was chosen to evaluate the cleanup options; the wastes were finally characterized as flammable materials, including tetrahydrafuran, methyl, ethyl, ketone, dioxane, and other acidic liquids. It was obvious now that the city could not perform the job by itself. Several public agencies were called to a meeting in April, 1980, to discuss the problem. Among those attending were the South Coast Air Quality Management Board, the County Health Department, the State Department of Health Services, the State Attorney General and the EPA. Boyer was also represented.

Sources of cleanup funds were identified. The agencies agreed to file law suits against the current property owner, the previous owner, Stankovich, and six companies believed to be generators of the material on the site. The suit, filed on May 14, 1981, by the State Attorney General, was based on public nuisance and California hazardous material laws. It requested preliminary and permanent injunctive relief. At the same time, the Attorney General was meeting with generators of the hazardous materials, including the Inmont Corporation of New Jersey, to negotiate funds for a cleanup.

Throughout, Stankovich maintained that the site did not violate laws and that the wastes were properly stored.

On July 10, 1981, a person was seen throwing a Molotov cocktail into the storage site. Some 120 fire fighters and six fire agencies responded to the ensuing inferno. Fireballs of gas erupted 200 feet into the air. Twelve thousand gallons of water per minute were sprayed on the fire, which took five days to extinguish.

Runoff from the fire fight carried chemical contaminants into a nearby creek. The State Department of Fish and Game estimated that the toxic liquid killed about 250,000 fish. A nearby bathing beach was closed for a day as a precaution. Prevailing winds carried air contaminants toward a neighboring city.

More meetings of state and federal officials and the generators of hazardous materials followed. A cleanup was finally completed in January, 1982. The site was scraped and six inches of top soil were removed. A clay cap was laid over the property. Compliance was finally achieved by changing the site to a burial ground. The price tag on the cleanup, $2,250,000, was shared by Inmont, $1 million; Superfund, $1 million; and the city, $250,000.

Both the State of California and the EPA brought actions against Boyer to recover costs of the clean up. In turn, Boyer sued the city for any costs which he

[37] *Id.*

is legally required to bear. Boyer negotiated a settlement in the state action. He agreed to undertake special measures to address concerns over the long term use of the property. As of this writing, the site is undergoing review for development as an industrial use.

About the resolution of the case, NRDC staff attorney Jonathan Lash later testified: the "private settlement . . . was widely viewed as much too soft. The settlement not only limited the company's cleanup responsibility but committed EPA to testify on behalf of the company in any subsequent lawsuit against it arising from the dump and the fire."[38]

Case 4: An Automaker Is Charged with a Technical Violation

It was a tough issue, a case of first impression. We had to bring all our persuasive powers to bear. I would not say it was a simple legal analysis.[39]

On December 8, 1976, EPA Administrator Russell Train wrote John Riccardo, President of the Chrysler Corporation, instructing the company to submit to the EPA a plan for bringing nonconforming automobiles into compliance with federal air quality regulations, specifically, with the Clean Air Act interim carbon monoxide standard of fifteen grams per mile. The Train letter began a case which involved alleged design deficiencies in the emission control systems of certain Chrysler Corporation automobiles.[40]

Chrysler used a carburetor idle, an alternative to the more expensive oxygen pump method for achieving adequate oxygen–fuel mixtures.[41] The government's position was that the system was so complex that it was forseeable that it would not be maintained properly in ordinary mechanics' shops; an infrared analyzer, a device not readily available, was needed for calibration. The government asserted that even a minor turn of a screw could change the carburetor idle fuel mixture and increase emissions up to 23 times Chrysler's specifications.[42]

---

[38] Written testimony by J. Lash. Hearings before the Committee on Environment and Public Works, United States Senate, 98th Cong., First Sess., Feb. 15, 1983, 46–50 at 265. Washington, DC: United States Government Printing Office, 1983.

As a result of information gathered in this affair and of inspections of two other sites which Stankovich occupied, a criminal action was brought against Stankovich. He was charged with violations of California hazardous waste statutes, and was convicted and sentenced to serve three years in prison. Telephone conversation with Fred Latham, Assistant City Manager, Santa Fe Springs, Sept. 17, 1985.

[39] Interview with Angus MacBeth, Feb. 2, 1985.

[40] *Chrysler Corporation v. EPA,* 14 ERC 1647.

[41] Chrysler had equipped some cars with air pumps in 1973 for comparison purposes. *Brief for Respondents, Chrysler Corporation v. EPA,* at 23, n.31 (hereinafter cited as Respondents' Brief).

[42] *Chrysler Corporation v. EPA,* 14 ERC 1647, 1652 (fn. omitted).

Cars in the recall class experienced problems in driveability and idle roughness; and warranty repair records indicated that 144, 891 claims had been submitted with respect to the carburetor during the warranty period.[43] Furthermore, Chrysler's reimbursement schedules for mechanics at Chrysler dealerships did not reflect the amount of work required to complete the adjustment task. When the work was completed the consumer was often dissatisfied with automobile performance. One testing program concluded that, in service, 90% of the affected automobiles would exceed the federal carbon monoxide standard.[44]

EPA's recall order applied to 208,000 1975 model automobiles, one fourth of that year's production line. These were Chrysler Cordobas and Newports, Plymouth Furies and Grand Furies, Dodge Monacos, Charger SEs and Coronets with 360 and 400 cubic centimeter engines equipped with two-barrel carburetors and catalytic converters.

The government argued that "Chrysler . . . designed a system which, as cheaply as possible, would provide barely minimal compliance . . . these cars do not in fact comply."[45] Furthermore, compliance required not only initial capacity to meet the standards but capacity over time: the system had to be designed reasonably to allow ongoing compliance. The EPA monitors automobile compliance to the Clean Air Act[46] in three stages: (1) examining prototypes of the new automobile line and issuing certificates of conformity to vehicles which pass the test, (2) testing a sample of vehicles after they come off the assembly line, and (3) testing the automobiles in actual use. Chrysler had passed the first two tests.

The government made its case on a voluminous record and argued that the company's violations of the Clean Air Act were major. In its brief, the government concluded that Chrysler's interpretation of compliance with the act "distorted the statutory goal, and the statute's structure, language and history."[47] The government emphasized in the judicial proceedings that the administrative judge had concluded that Chrysler's misadjustments were the "inevitable by-product of their emission design and service procedures."[48]

The agency's conclusion about the adjustment process was summarized on review in federal court.[49]

---

[43] *In the Matter of Chrysler Corporation*, United States Environmental Protection Agency, Before the Administrator, pp. 99–100.

[44] This was the figure, before maintenance, reached in the Olson Program. Other figures ranged from 25% to 100% of vehicles that were out of compliance.

[45] Respondents' Brief, at 59.

[46] 42 U.S.C. § 202(a)(1), (d)(1), 42 U.S.C. § 7541(a)(1), (d)(1).

[47] Respondent's Brief, at 35.

[48] *Chrysler Corporation v. EPA*, 631 F. 2d 865, 893.

[49] *Chrysler Corporation v. EPA*, 14 ERC 1647, 1657 (quoting JA 111:912–913).

The adjustment process is cumbersome and time-consuming, taking the mechanic approximately 30 to 40 minutes, according to an EPA investigation. The mechanic must first obtain a Chrysler Huntsville exhaust emission analyzer or other approved infrared analyzer, and must verify that it is warmed up and calibrated according to the manufacturer's instructions. He must check to see that the sample lines and connections of the sampling system for the analyzer are free of leaks and must also warm up the vehicle's engine and allow it to idle for no more than ten minutes. Then the mechanic must remove the plug from the threaded catalyst tap on the vehicle and install the sample line of the analyzer in the tap upstream of the catalyst. This will generally require the mechanic to crawl under the car or to use a hoist. He must then start up the vehicle's engine and verify that the idle RPM and timing are within specification. At this point he must measure the idle carbon monoxide concentration. If it exceeds specified limits he must adjust the mixture screws on the carburetor to achieve a "leaner" mixture of fuel and air in the idle circuit, checking back and forth between the analyzer and the carburetor to monitor the effect of his adjustment on the idle mixture. A clockwise turn of a screw will decrease the amount of fuel discharged through the idle port and into the idle circuit; a counterclockwise turn will increase the amount of fuel, resulting in a "richer" mixture. The adjustment screws are highly sensitive and must be turned in small fractions of a rotation in order to set the adjustment with sufficient precision. After making the proper adjustment the mechanic must also balance the mixture screws on a two-barrel carburetor for lowest level of hydrocarbons or smoothest idle within the prescribed specifications (footnote omitted).

Administrator Russell Train concluded:

Chrysler is responsible for these misadjustments because Chrysler as an automobile manufacturer should have foreseen that its carburetor design and adjustment procedures would cause widespread misadjustments and because of the agency relationship which exists between Chrysler and its authorized dealerships.[50]

Chrysler's view was quite different. The company stated that "E.P.A. agrees that our cars meet the Federal standards when maintained to specifications. . . . However, when an individual improperly adjusts an emissions control system we are not responsible for those actions."[51] The company charged that the EPA was "trying to require Chrysler to be responsible for the actions of private individuals."[52] For the first time an automaker decided to contest a government recall order.

Chrysler characterized the adjustment procedure as "relatively simple,"[53] able to be "done by a normal guy with normal equipment";[54] it is "easy for a mechanic to make a quick and precise adjustment of the idle mixture."[55]

Chrysler addressed the problem and the recall aggressively. The company

[50] *Chrysler Corporation v. EPA,* 14 ERC 1647, 1651.
[51] *The New York Times,* Dec. 11, 1976, at A–1.
[52] *The Wall Street Journal,* Dec. 13, 1976, A–4, col. 2.
[53] *Chrysler Corporation v. EPA,* 14 ERC 1647, at 1651, n.21 (Freer statement).
[54] *Chrysler Corporation v. EPA,* 14 ERC 1647, at 1651, n.21 (Heinen statement).
[55] *Chrysler Corporation v. EPA,* 14 ERC 1647, at 1651, n.21 (Brubacher testimony).

did not submit a recall plan; rather, it pursued administrative and legal procedures. The corporation was concerned about the precedent it felt the government was trying to establish.[56] It challenged the government's testing procedures, argued that the misadjustment process was not foreseeable and that Chrysler dealers were not agents of the corporation in performing adjustment services. Further, the company argued that a recall was based on retroactive rule making and that the language of the Clean Air Act was ambiguous; the act

> was written in such a way to cause enormous problems. . . . (It) essentially wrote out any consideration of whether you're getting your money's worth.[57]

The specific section of the Clean Air Act under consideration was 207(c), which states in part:

> If the Administration determines that a substantial number of any class or category of vehicles or engines, *although properly maintained and used,* do not conform to the regulations prescribed when in actual use throughout their useful life . . ., he shall . . . notify the manufacturer.

Chrysler argued that the italicized language meant "maintained in accordance with the manufacturer's specifications." The government concluded that the violations of the Clean Air Act were several and involved Section 207 (c)(1): improper maintainance is no defense for violations of the act where the manufacturer is itself responsible for the violation. Also, both the owners and mechanics did follow the manufacturers instructions and Chrysler could not legally defend its actions by arguing noncompliance by either of these groups. In the face of the magnitude of the violations, whereby up to 79% of the implicated cars were violating the carbon monoxide standards, Chrysler's views "made a sham" out of the Clean Air Act provision.[58] The EPA administrative law judge took a compromise position concluding that the phrase means "maintained in accordance with the manufacturer's (instructions or) specifications" and that "EPA's interpretation . . . is contrary to the ordinary meaning placed upon it both by the industry and the 'reasonable man' theory."[59] Nonetheless, he placed the burden of persuasion thus: "EPA, a third party to the maintenance of the vehicle, need only show . . . substantial compliance with the manufacturer's instructions."[60]

The reviewing court responded to the ambiguity argument delicately:

---

[56] Chrysler interview, Aug. 27, 1981.

[57] *Id.*

[58] Respondents' Brief at 3–5.

[59] *In the Matter of Chrysler Corporation,* United States Environmental Protection Agency, Before the Administrator, at 14.

[60] *Id.,* at 12.

> We agree that Chrysler has proposed one plausible interpretation of the Act, but we
> cannot agree that no other interpretation is possible. . . . The difference in language
> may indicate a subtle difference in meaning.[61]

Chrysler concluded that consumer behavior could cause the problem and that the auto manufacturer should not be responsible "if the owners failed to attain maintenance with the manufacturer's precise written specifications." Chrysler further asserted that the government should seek compliance "by suing the mechanics for tampering."[62] A requirement of future changes in the technology, which Chrysler had planned, not a recall, "would be a much better use of resources."[63] But the Clean Air Act stated that the proper remedy was recall of the entire class of vehicles: "Nonconformity . . . will be remedied at the expense of the manufacturer."[64]

The case was considered not only wasteful but silly by the company:

> From the broadest perspective, there were a few cars violating a CO standard. In terms
> of air quality and public health there was no consequence whatsoever from these
> technical violations. Obviously we were exceeding the standard but we have to ask
> what should the role of government be. What were they [EPA and the court] supposed
> to be deciding most basically?[65]

The government position on health effects was quite different. EPA recognized CO as a poisonous gas that at levels of exposure being experienced in several American cities negatively affected, among other things, mental acuity.[66]

Chrysler, in contesting the government allegations and conclusions, argued that "most analyzers were in operating condition and working with sufficient accuracy for the use for which they were intended."[67] The adjustment process was not difficult; it merely took "slightly longer" than other procedures.[68] The company contended that federal standards would be exceeded if between 1.5 and 2.0% carbon monoxide concentrations were found in the tail pipe; the government administrative law judge (ALJ) put the figure at 1.0%.[69] The ALJ found that data from testing programs indicated that a requisite "substantial number"

---

[61] *Chrysler Corporation v. EPA*, 14 ERC 1647, 1665.

[62] Respondents' Brief, at 71.

[63] Chrysler interview, Aug. 27, 1981.

[64] 42 U.S.C. §§7541(c)(1), 207(c)(1).

[65] Chrysler interview, Aug. 27, 1981.

[66] Specifically, EPA, citing a 1968 HEW report, legislative history of the Clean Air Act and reports of the Council on Environmental Quality, noted that carbon monoxide impairs the oxygen carrying ability of the blood; that it can affect visual acuity, motor ability and mental performance in levels found in cities and that its effects can be severe. Respondents' Brief, at 3–5.

[67] *Chrysler Corporation v. EPA*, 14 ERC 1647, at 1652, n.31 (quoting JA V:2001). Memorandum from A. T. Weibel to W. S. Fagley.

[68] *Chrysler Corporation v. EPA*, 14 ERC 1647, at 1652 (quoting Brief of Petitioner at 65).

[69] *Chrysler Corporation v. EPA*, 14 ERC 1647, at 1654, n.56.

of vehicles did not comply. Chrysler argued that EPA needed to show that 50% or more of the recalled group was violating the standard.[70] The company also attacked the government's statistical sampling approach.

Chrysler made its case before the ALJ, before the new EPA administrator, Douglas Costle, and in the federal courts. Each level upheld the government.[71]

Chrysler claimed that the case, like many suits brought by the EPA during the Carter administration, demonstrated an unbalanced environmental policy. One frustrated company spokesman summarized the company's position:

> In the whole Clean Air Act what's written is that we've got to protect, not only healthy people, but the asthmatics and so on. Maybe it'd be cheaper just to put them all in a hotel room for the rest of their life . . . . Filter in the air. I'm not suggesting this, but . . . .[72]

The company considered the case an unfair attack by environmentalists within government and an improper legal interpretation. The court relied on statutory history that was irrelevant to the recall issue and—at various levels of consideration—a theory of products liability not found in the statute. A law review commentary some years later concluded:

> While manufacturers undoubtedly *should* strive for design free of defects, there is no indication in the statute, the legislative history, or cases decided under other sections of the Clean Air Act that Congress intended to impose any form of strict liability upon automobile manufacturers. Thus, products liability doctrine alone cannot support the court's interpretation of "properly maintained and used."[73]

Chrysler executives felt that the nature of the violation should be considered within a context of the regulatory demands put on the automobile industry during this period of implementation of the Clean Air Act.

> I think the problem really was one that was difficult to foresee . . . We were dealing with a substantial increase in the severity of the standards from '74 to '75—on a very short time frame. We had to incorporate catalytic technology in vehicles, very new technology. There was an awful lot of work that went into that.
>
> We determined after the fact that some of the existing vehicle technology might have had some changes which might have improved the in-use performance of those vehicles. There have been changes to that technology since then that do not eliminate the 1975 type of problem but certainly reduce it. I guess from my standpoint that is a problem that would be better handled, instead of forcing a manufacturer to recall a

---

[70] *Chrysler Corporation v. EPA,* 14 ERC 1647, 1659, n.84.

[71] The case was finally concluded with the United States Supreme Court denying a writ of *certiorari,* 49 LW 3410 (Dec. 2, 1980).

[72] Chrysler interview, Aug. 27, 1981.

[73] Comment, "*Chrysler Corporation v. EPA* , 631 F.2d 865 D.C. Cir. 1980," 11 *Environmental Law* 784 (1981).

series of vehicles, to make some kind of change. I think it would be better to make those changes in the future. I think it would be a much better use of resources.[74]

The government saw the options in a much narrower perspective:

An agency has a very real problem if it attempts to adopt a position not taken by Congress. It was remedially bound by statute. The short answer from government is that to adopt that remedy would go beyond statutory instructions. . . . That's not to say it (a more flexible response) isn't a good idea.[75]

[74] Chrysler interview, Aug. 27, 1981.
[75] Interview with Angus MacBeth, Feb. 2, 1985.

CHAPTER 2

THE NONCOMPLIANCE PROBLEM

Introduction

The *Frezzo, Distler, Inmont,* and *Chrysler* cases help introduce the problem of gaining compliance with the nation's environmental laws.

James and Guido Frezzo's experience raises troublesome issues. Is the small businessman an appropriate target of prosecution? Among all violators, is he unfairly selected? Are criminal sanctions the best means of altering the behavior of people who violate regulatory law, especially when the alleged damage is minimal? Is education of polluters as important as their punishment? The *Distler* case demonstrates that some violations clearly are *mal in se* and merit the full power of a coordinated criminal justice system. *Distler* also introduces the challenging tasks of employing scientific evidence in environmental legal proceedings and, more generally, of establishing causation as a basis for environmental regulation. The *Inmont* affair underscores the difficulty of achieving compliance when many possible culpable actors and many sources of diffuse government power exist and few assume responsibility for clean up. *Chrysler* highlights fragmented responsibility in another way: the consumer, the government, and the industrial giant all played roles in designing and implementing a pollution control system sure to fail.

These are four of hundreds of cases of violation of state and federal pollution control law. We will return to them in developing an understanding of compliance, in presenting a framework for promoting compliance, and in suggesting policies that can improve the implementation of environmental law.

First, however, it is important to become familiar with the general problem. What is the size of the problem? Who is violating the nation's environmental laws? What are they doing? Is what they are doing serious? Why is a consensus on environmental compliance so elusive?

The Nature of Environmental Violations

The problem of violations of environmental law is immense. A few statistics give an indication of its size. In fiscal years 1977 through 1980 EPA enforcement actions for nonmunicipal sources allegedly violating National Pollution Discharge Elimination System permits (NPDES), the kind James Frezzo should have sought, numbered 2,366;[1] and 424 notices of noncompliance of the Toxic Substances Control Act were issued.[2] In 1983, 60% of facilities required to undertake groundwater monitoring were out of compliance.[3] In one Southern California air quality management district alone, in one representative year, 10,500 complaints of noncompliance were received. On the basis of these and the district's self-initiated inspections, which numbered about 50,000, the district issued 6,000 notices to comply. These were followed by 2,500 notices of violation of air quality rules targeted to facilities ranging from refineries and power plants to service stations and dry cleaners.[4] Many nuclear power plants have not complied with Nuclear Regulatory Commission (NRC) rules. Forty-two of forty-eight plants missed the July, 1980, deadlines for installing accident warning systems. As late as February, 1982, a fifth of the plants still had not complied.[5]

Noncompliance occurs in all economic sectors. While no exact statistics exist, a single individual or small business is responsible for a large percentage of violations. Some activities are criminal; others, civil wrongs or infractions.

The violations involve a considerable range of behaviors. A recent inventory includes: falsifying of compliance reports to EPA;[6] illegal storing of toxic wastes;[7] conspiracy to falsify manifests[8] and to dispose of wastes along rural roadsides illegally;[9] conspiracy to bribe officials to allow dumping of toxic materials into ordinary landfills;[10] entering into ''secrecy pacts'' in which the buyer of a toxic material is required to assure the selling company in writing that materials will not be traced back to the seller; and adding chemicals to drums

---

[1] This included notices of violations, administrative orders, and referrals to the United States Department of Justice (EPA, Office of Water Enforcement). Environmental Quality Council, 12th Annual Report, *Environmental Quality, 1981.*

[2] EPA, Office of Pesticides, and Toxic Substances, Environmental Quality Council, 12th Annual Report, *Environmental Quality, 1981.*

[3] *Inside EPA,* Apr. 8, 1983, at 14.

[4] The South Coast Air Quality Management District: A Progress Report, 1977–1983.

[5] *Los Angeles Times,* Feb. 2, 1982, at 2, cols. 5–6.

[6] *Nugent and Matula.* (Throughout the footnotes cases assembled for the study are referred to by a short name. Complete citations are found in the appendix. Other citations are given fully in the notes)

[7] A. C. Lawrence Leuther case reported in *Rolling Stone* 10 (Mar. 29, 1984).

[8] *Nugent and Matula.*

[9] *Burns* and *Ward.*

[10] *The New York Times,* Nov. 19, 1982, at B2, Col. 4.

containing toxic wastes to disguise odors which would identify waste substances requiring special handling.[11] Further, the direct disposal of highly toxic, even lethal, wastes into municipal sewer systems and onto the land such as into coal mines has been increasing.[12] Illegal sales of waste oils contaminated with toxic substances; dredging and filling of waterways without a permit;[13] improper filling of wetlands;[14] and the unlawful taking of marine mammals[15] are chronicled.

Many actions by small business are inadvertent or negligent or otherwise lack criminal intent. These include failure to secure one of several discharge permits, to provide complete information, or to maintain control technology.[16] A farmer may ignore a leaky holding pond for too long; a dry-cleaning firm may forget to check filters according to suggested schedules.

Wanton and willful violations by organized crime in businesses subject to environmental regulations are at the other end of the continuum from the inadvertent violation. The extent of the problem is unclear. Studies differ on how to assess the degree of involvement by criminal rings and resulting violations. Many charges brought by government officials are vague and undocumented, for example, "infiltration" into the waste disposal industry and "surreptitious" dumping. As Congressman Albert Gore observed in congressional hearings:

> The fourth allegation was that SCA [the company under Congressional scrutiny] had acquired a large number of companies in New Jersey and that some of those companies were controlled by figures associated with organized crime—and that word "associated" is a very difficult and tricky one but one that police investigators have been forced to resort to in the investigation of organized crime in whitecollar pursuits.[17]

Organized crime is most often implicated in improper disposal of toxic wastes which offers tempting profits. Indeed, the "take" can be staggering. Recently the State of Pennsylvania successfully prosecuted a businessman under felony racketeering charges. William A. Lavelle was accused of profiting from illegal disposal of toxic materials into abandoned anthracite mines under Scranton and Wilkes-Barre. According to the government, Lavelle earned $580,989 from 1976 to 1979.[18]

---

[11] *The New York Times*, Feb. 21, 1983, at A12, cols. 1–2.

[12] *Aaro, Culligan, Custom Plating, Distler, Lavelle.*

[13] *United States v. M.C.C. of Florida, Inc.,* No. 81-2373-EBD (S.D. Fla., Dec. 17, 1982), 13 ELR 20305.

[14] *United States v. Carter,* Nos. 81-981-Civ-JWK (S.D. Fla., Dec. 21, 1982), 13 ELR 20307.

[15] *United States v. F/V Repulse,* No. 81-3182 (9th Cir. Sept. 28, 1982), 13 ELR 20554.

[16] *Frezzo Brothers.*

[17] U.S. Congress, House of Representatives, Committee on Energy and Commerce, "Organized Crime Links to the Waste Disposal Industry." Hearings before Subcommittee on Oversight and Investigations, May 28, 1981, at 33. Washington, DC: Government Printing Office, 1981.

[18] *The New York Times,* July 18, 1983, at A6, cols. 1–4; *The Wall Street Journal*, July 8, 1983, at A8, col. 6; *Los Angeles Times,* Apr. 25, 1984.

The variety of environmental violations by American industry is striking. Cases have involved almost every conceivable deficiency in a production, disposal, or transportation process. They include improper construction of toxic waste disposal sites,[19] receiving substances not permitted for a site,[20] deliberately failing to report to government authorities, and destroying test results.[21] Government has also prosecuted pollution of rivers and streams with deadly chemicals;[22] improper and illegal fuel switching in a corporate automobile fleet;[23] the Chrysler type of violation, that is, improper design of emission control equipment;[24] unauthorized maintenance of engines used for emission control tests;[25] withholding test information from government;[26] illegal disposal of pollutants into waterways;[27] nonregistration or misbranding of pesticides or toxic substances;[28] illegal shipping of hazardous wastes across international boundaries;[29] improper housekeeping practices resulting in pollution overflows from holding tanks into waterways;[30] improper removal from power plants of seemingly minor parts such as fuses;[31] improper securing of facility fences in utility plants;[32] and poor construction practices and harassment of control personnel at nuclear plants.[33]

Extensive noncompliance results from actions of government agencies themselves. Our focus in this book is mainly on private sector violations. Getting public agencies and supposedly subordinate levels of government to comply with laws and judicial decisions is a major issue in environmental law. Some of the present analysis will clearly generalize to compliance in the public sector. However, our data come from investigations of the firm. In addition, some factors in the framework presented in this book may differ when government is a target. For example, within government status differences between regulators and the regulated may be less pronounced. Second, the nature of the regulatory process has more commonly been defined as adversarial in the business–government

[19] Waste Management Co.; *The Wall Street Journal*, Jan. 6, 1984, at B37, col. 2.

[20] *Id.*

[21] *Id.*

[22] *Allied;* Corning Fibers Corporation, *Los Angeles Times*, Nov. 27, 1981, at 2, cols. 6–7.

[23] *Arco.*

[24] Chrysler.

[25] Ford; *The New York Times*, Feb. 9, 1977.

[26] *Dow.* See fn. 17, Chap. 3.

[27] *Kennecott,* Mobil, *Los Angeles Times*, Sept. 28, 1982.

[28] *In re U.S. Polychem Corp.*, Environmental Protection Agency, Notices of Judgment Under FIFRA, June 1975, No. 1466 (May 3, 1974).

[29] *Matula and Nugent.*

[30] Aluminum Company of America.

[31] *Virginia Electric & Power Co.*

[32] *Tennessee Valley Authority.*

[33] *The New York Times*, Dec. 2, 1981, at 1, col. 3.

context. Finally, the organizational cultures of government agencies may be more similar than when government is compared with the private enterprise.

Nonetheless, the noncomplying acts of government must be noted because they add to the context in which business will respond to law and because government is often directly responsible for the illegal behavior of the firm. Municipalities fail to comply by behaviors that range from disposing of materials in violation of federal standards to removing emission control devices on their police cars.[34] The General Accounting Office found that in one year, 1980, at least 43% of the country's 65,000 community water systems did not comply with federal safe drinking water standards; over 146,000 violations were noted.[35] A 1978 EPA study reported that only 35% of the country's 14,000 municipal landfills were in compliance with state regulations,[36] and a 1983 investigation concluded that over 60% of the facilities surveyed did not comply with ground-water monitoring requirements.[37] Government fails to follow procedural require-ments such as under the National Environmental Policy Act[38] and fails to meet legislatively imposed deadlines to promulgate rules.[39]

Government is also implicated in ways that affect business compliance. Agencies may give erroneous guidelines to businesses which then are sued through initiatives of other agencies, or by private citizens or public interest groups. For example, General Motors found itself implicated in a suit because it followed EPA directives to offset automobile emissions with reductions in future models; in fact, the agency was legally required to recall 1979 Pontiacs that did not meet federal requirements.[40] Or one branch of government may fail to impose a requirement on industry which leads to vulnerability to civil actions. Government may fail to provide funding or leadership which is necessary to implement a pollution control program in a timely manner. Or government can promulgate rules that are arbitrary and capricious and thereby create pressure for businesses to fail to comply.[41]

Official improprieties can lead to noncompliance. A Teamsters Union suit alleged that EPA officials had described to a petitioner how discretionary ap-proval of a disposal facility could be obtained, although approval would violate

[34] *The New York Times,* Nov. 24, 1983, at A–16, col. 6.

[35] *Inside EPA,* Mar. 12, 1982, at 10.

[36] Environmental Quality Council, 11th Annual Report, *Environmental Quality 1980.*

[37] *Inside EPA,* Apr. 4, 1983, at 14.

[38] N.E.P.A., 42 U.S.C. §§ 4321–4370. In 1980 alone 140 NEPA cases were reported by federal agencies; most were brought by individuals or citizen groups. Seventeen of these cases resulted in an injunction. Environmental Quality Council 12th Annual Report, *Environmental Quality,* 1981.

[39] Failure to promulgate premanufacturing rules under the Toxic Substances Control Act 15 U.S.C. §§ 2601–2629, is an example.

[40] Civ. A. No. 82-2910, 558 F. Supp. 103 (D.D.C., Mar. 2, 1983), 13 ELR 20752 is an example.

[41] *Western Oil and Gas Association v. California State Air Resources Board,* 2 Civil No. 63339 (Cal. Ct. App. Mar. 10, 1982), 13 ELR 20447.

provisions of both the Toxic Substance Control Act and the Resources Conservation and Recovery Act.[42]

For other cases, it is difficult to ascertain the exact nature of the noncomplying act or of the responsible party or parties. *Inmont* falls into this class of
violation. Hundreds of sites contain thousands of barrels of chemical wastes;
stored barrels corrode and chemicals leak onto the land and into the water table.
EPA has ranked sites in terms of potential danger to health and environment. But
often the contents of the barrels, the owners of the materials, and the original
waste haulers are unknown. In some cases, it is not clear who owns the storage
site. A regional water quality control board in California had to resort to a
multiple liability theory in a case wherein it could not determine who was most
responsible for noncompliance, that is, who caused ground water contamination
by leaking trichlorethylene. One official concluded: "We feel they're all part of
it. . . . They all have leaks, and we can't sort out who lost five gallons and who
lost 100 gallons." Each company was ordered to contribute to the clean up.[43]

Exxon was involved in a controversy in 1984 which did not go to trial and
which is difficult to classify in a noncompliance typology. The company had
been shipping millions of gallons of freshwater from the Hudson River to, among
other sites, a Caribbean island. Company ships would enter the Hudson, unload
salt water, take on fresh water and carry their new cargo to overseas process
operations. Outraged environmental groups sought to litigate. Although the
nature of the violation was not clear, Exxon agreed to stop and paid $500,000 to
the Hudson River Fisherman's Association and the Open Space Institute. The
New York State Legislature began work on a law to regulate the "theft of
water."[44]

Other problems result from acts of those who are intruders at a pollution
source. Vandals caused two major toxic spills into the Russian River in northern
California; 21,000 gallons of formaldehyde ran into the river.

Not every harmful act constitues noncompliance, despite public belief.
Temporary and minor problems occur that are not covered by any regulation nor
by common law causes of action such as nuisance. Occasionally, a potentially
disastrous act may result from beneficence. The dangerous spread of toxic materials in the Missouri dioxin case was exacerbated when the St. Louis Police
Academy allowed local gardeners to scoop up free manure later found to contain
dioxin.[45]

Noncompliance, thus, is an immense problem in all sectors of the economy.
Violations result from the acts of innocent individuals and from the refined plans

[42] *The Wall Street Journal,* Apr. 22, 1983, at A4, cols. 4–6.
[43] *Los Angeles Times,* June 18, 1984, at 2, col. 6.
[44] *Los Angeles Times.*
[45] *The New York Times,* Feb. 24, 1983, at B14.

of sophisticated criminal groups. Government, responsible for enforcement and for compliance, is often involved in violations. Policies developed to promote compliance confront a range of behaviors of agencies and firms.

The Nature of Compliance

Describing and quantifying alleged and actual violations of environmental regulations is but one method of addressing noncompliance. Such counting is not notably useful for resolving public policy questions raised by environmental law—questions such as those related to the definition of compliance and the importance of achieving compliance. Different views of these two issues explain much of the rift between government and business regarding environmental controls.

What Does It Mean to Comply?

This deceptively simple question is answered in several ways. Interest groups promote different interpretations because of the significant financial, environmental, and philosophical implications of deciding what it means to comply. They can do so because the legal system develops a tolerant context. Lawmakers often do not specify the meaning of words and phrases. They fail to give detailed interpretations—sometimes intentionally, sometimes on the basis of oversight, and sometimes because they lack necessary background information. People with very different technical and disciplinary backgrounds—lawyers, engineers, accountants, public administrators—then read and interpret laws and regulations. These groups employ different means to calibrate compliance. Standard operating procedures in large organizations also produce different meanings of words, no matter what their environmental objectives. It is not clear, then, whose definitions to use.

A basic distinction exists between specific compliance and general compliance. Specific compliance refers to a response of a business targeted by an incentive or sanction that is believed to be consistent with societal objectives or regulations. General compliance refers to responsive behavior of the universe of businesses whose law-abiding performance government aims to effect. Another important concern is whose definition of compliance to employ. Determinations about violations and law-abidingness often are left to the lawyers, those skilled in the processes of aligning meanings in order to fit their clients' interests. How does one define the "best practicable technology" (BPT) as used in the Clean Water Act? In achieving such a standard, does one also meet the criteria of "best achievable technology" (BAT) required by 1984? The Chemical Manufacturers Association has maintained that BPT is effective enough to reach environmental goals; therefore, it can be classified as BAT.[46] What is a "feasible and prudent alternative"? Does it include the alternative of no action? What is an "environmental impact"? Does it include psychological effects on vulnerable populations from projects in the urban environment?[47] What is a "significant environmental impact"? What is an unnecessary risk? What is a "properly maintained" emission control system?

Words have ambiguous meanings in most contexts. In the world of legislation, those concerned with government policy on the environment strive to introduce their interpretations into statutes, regulations, and judicial opinions or to confuse definitions in order to allow greater flexibility in pursuing interests. Ambiguity is often sought.

Definitional choices carry important financial and administrative implications for affected companies and for government. Some options can have major environmental effects, some are the stuff of black humor, and some are both. The semantic sleight of hand that can enter into presumed compliance is exemplified by a Federal Trade Commission action against the R. J. Reynolds Company. In 1982, the company agreed to pay civil penalties of $100,000 for practices that it had earlier insisted were in compliance with federal rules. These involved the inclusion of health warnings in cigarette advertising. The company first said it had complied with the rule when it printed the warning in a language different from that of the rest of the advertisement.[48]

Issues of interpretation are not limited to legal analysis. Some debates about compliance are statistical. For example, an industry analysis of an EPA groundwater monitoring study asserted that "serious statistical flaws . . . [would] erroneously show 80% of facilities are violating ground water requirements by the end of the first year of monitoring and that every facility in the country [would be] violating the standards by the end of the second year."[49] By one method of statistical analysis, compliance may be demonstrated, by another, a violation found. An EPA policy document identified ten different "yardsticks" for establishing compliance with water quality regulations. They included the ratio of the population served by a municipal wastewater treatment system to the total population that should be served by treated water as well as the extent of the return of valuable fish to previously polluted streams.[50] These examples underscore the importance to compliance evaluations of the identity of the "environmental accountants," those persons who apply rules to the evidence.

Compliance is an ongoing process; therefore, timing of the determination is

---

[46] *Inside EPA*, May 7, 1982, at 11. See also Reed, "New BAT Standards: Lowering The Ceiling or Raising the Floor?" *Environmental Law Reporter* 13 (January, 1983), 10002.

[47] *Metropolitan Edison Company v. People Against Nuclear Energy*, 103 S. Ct. 1556 (1983).

[48] *The Wall Street Journal*, July 14, 1982, at 22, col. 6.

[49] *Inside EPA*, Oct. 22, 1982, at 3.

[50] *Inside EPA*, Sept. 17, 1982, at 9.

critical. Typically, government and business will differ on how long industry should be given to come into compliance. They also are apt to disagree on the extent of monitoring to be used to determine whether there has been a violation. Should ongoing operation and maintenance problems in reaching standards be treated with the same significance as original failures to comply?

Thresholds of compliance also can be controversial. At what point should government conclude that an enforcement effort or other official response is required? It may be argued that no discretion and no interpretation should be involved. The government must act when there is an unambiguous violation. But environmental enforcement is much like many areas of law enforcement. It is impossible to detect all violations, and the government cannot afford to pursue all those which are detected. Good faith efforts, then, may be equated with compliance, or compliance with "most of the regulations, most of the time" may become the standard. Perhaps only those who fall into patterns of violations or those whose actions significantly affect the environment will be cited. The *de minimis* violation would thus prove of no interest.

Should an occasional failure to achieve a standard be categorized as non-compliance that merits government action? A rare runoff from an agricultural pond into a neighboring waterway, a single leaking barrel among thousands of stored barrels, a lone failure to record in a manifest system the transfer of a hazardous product from a producer to a disposer, an emission misfunction of a thousand out of a million cars, or of 50% or more of a recalled group of automobiles, as Chrysler argued in our fourth case? Should government consider these noncompliance?

Variations in the meaning given to compliance tap significant differences in philosophies regarding the goals of environmental programs and their relative place among other social aims. For some, the objectives of federal environmental laws are primarily symbolic: to pursue them literally would be totally unreasonable. Others believe that each regulation is a societal value statement; neither individual firms nor individual government administrators can implement or enforce the rules selectively. Some legal scholars reject the notion that regulatory law can be anything more than symbolic. Agencies "manufacture the appearance of activity . . . the symbolic reality of impact, the fiction of real power."[51] Modern-day regulation is attacked as being naively instrumental and based on models of policy implementation that are simplistic and wrong (Teubner, 1984). The regulated react with enthusiasm to this idea.

Government responses to this range of views can vary. Courts may devise notions of "substantial compliance" with the spirit of the law (Comment, 1982). Agencies may develop a concept of "tolerable noncompliance," especially

---

[51] Peter Manning, quoted in Keith Hawkins, *Environment and Enforcement: Regulation and the Social Definition of Pollution* (New York: Oxford University Press, 1984), at 10.

when important organizational goals are being reached (Viscusi & Zeckhauser, 1979). A certain level of stealing of government resources reportedly became the accepted norm in the Pentagon's relations with some suppliers. Military men were said to have a ''sort of conflict of interest. . . . Their main mission is to get the ships built, to get the meat purchased. There's an attitude that maybe there's a tolerable amount of stealing that's going on.''[52] Similar tolerance is found in many administrative agencies.

Agencies may promote procedures that regard full compliance as unimportant. Compliance is seen as an event, and the development of standards is viewed as an ongoing process which the regulated can shape. Certain government–business relationships are so intimate that adherence to rules means little more than that government blesses existing business activity. In analyzing enforcement of New York State's environmental laws, an assistant attorney general put the matter this way: ''Enforcement [has become] a process of whittling down the obligations of the polluter to the point where he can meet them.''[53] This relationship between government and industry can be acceptable, if not benign. The regulator needs information from the regulated to develop reasonable and meaningful rules. However, when goals are continuously altered in response to performance, compliance can become an empty concept. Conditions necessary to reform rules to achieve meaningful compliance are not common; they include vigilant monitoring by those opposed to this comfortable business–government relationship.

Finally, should the definition of compliance be exclusively legal? Business and government can be said to comply formally with rules according to a particular governmental administration. Yet groups within the population may disagree with the conclusion. Is there an ethical dimension of compliance that must be addressed? If so, who is the keeper of the pure environmental objective? Traditional means of analyzing legislative intent can be used, and presumably similar techniques could also be employed to define judicial or administrative intent. But legal responses alone will not satisfy those who read primacy of environmental quality into the inspiring language of the major legal initiatives of the environmental decade. In fact, the intervention of lawyers in the processes of environmental regulation and management is sometimes viewed as an obstacle to achieving meaningful compliance.

Is Compliance Desirable?

Whether seeking compliance with environmental law is always a desirable end is another analytical issue. For some violations the answer is obvious. When failures to comply lead directly to significant health, ecological, or property

[52] *The Wall Street Journal,* Jan. 17, 1983, at 14, col. 1.
[53] *The New York Times,* Feb. 5, 1983, at 35, col. 4.

damage, few would disagree with the goal of compliance. But in many areas of regulatory law, both effects and causal relationships are unclear and give rise to the need to think more fully about the impacts of demanding obedience to extensive regulatory controls. Are the rules rational? Are they economically sound? Are they understandable? Does compliance further environmental objectives or is the infrastructure of regulation—the large bureaucracies and complex codes—independent of environmental quality?

To be sure, most Americans link noncompliance with extensive damage to health and the environment. Such links are more and more often discussed and understood. Polls reaffirm popular support of stringent environmental regulations[54] even when respondents are asked to consider the economic costs of pollution control efforts. The public feels that environmental laws must be maintained "in order to preserve the environment for future generations,"[55] and that environmental standards and requirements must be stringent. A federal study involving 60,000 citizens found that environmental crimes were ranked as very serious by the public. Most serious on the list of crimes described in the research was a bombing of a public building, but an illegal dumping that pollutes water supply and kills people was ranked as more serious than heroin smuggling and skyjacking.[56]

Of course, industry has its favorite statistics. For example, 50% of respondents to one recent survey felt that "government regulations dealing with energy . . . (have) . . . a great deal of impact on economic groups." Only 13% said that there was "no" or "not much" impact, and almost half of the respondents "felt the government pays 'too little' attention to the impact of regulations on economic growth."[57] And even studies which find strong support for pollution control also find a softening of the response when a trade-off with energy supply is posited.[58]

Some citizens conclude that compliance is a social good independent of the relationship between law and environmental quality. Whatever the nation's laws, they must be obeyed. Social order will be threatened and the fabric of democracy will unravel if the law is flaunted. Even when flawed, law must guide: it represents society's best efforts to address significant collective challenges. Individuals and individual businesses cannot decide the wisdom of law and the need for differential compliance.

---

[54] For example, see *The New York Times* and CBS News Poll, *The New York Times,* Oct. 4, 1981, at 30, cols. 3–6. See also L. M. Lake, "The Environmental Mandate: Activities and the Electorate," *Political Science Quarterly,* 2 (1983), at 215–233.

[55] *The New York Times* and CBS Poll.

[56] William Greider, "Fines Aren't Enough: Send Corporate Polluters to Jail," *Rolling Stone,* 10 (March 29, 1984), at 9–10.

[57] Cambridge Report no. 85.

[58] *The New York Times* and CBS Poll, n. 45 *supra.*

Yet, in light of the revolutionary outpouring of regulation (Mitnick, 1980) it is essential to treat seriously the analysis of the advisability of achieving compliance. Costs of pursuing compliance are high; the relationship between rules and environmental protection is sometimes unclear; and the health effects of increased control are uncertain.

Industry, which estimates that compliance will cost in the billions of dollars, complains about the expense of moving to the next increment of environmental quality. The EPA estimates that the cost to comply with the Clean Water Act will be $118.4 billion by the year 2000.[59] Individual industrial sectors have given equally staggering figures.

The Business Roundtable (Gatti, 1981) concluded that the incremental costs of government regulation to 48 companies participating in its study (in which *incremental* was defined as costs incurred for activities over and above what a company would have done without regulation) was $2.6 billion. The main share of these costs ($2 billion or 77% of the total) was attributable to EPA regulations; $36 million of this figure was incurred in maintaining environmental programs and monitoring regulations and proposed regulations. Within the EPA figure, 60% of the costs was attributable to the air program and over 30% to the water program. The motor vehicle emission and NAAQS particulates regulations represented $300 million of incremental costs.

Industry cites environmental regulation as the source of countless evils, including decreased productivity and increased unemployment due to the need to close down factories. And regulation is characterized as irrationally designed and unfairly and inefficiently enforced. The Conference Board (Lund, 1977) has described what it believes to be several major problems with the nation's environmental laws: overregulation, conflicts in and overlap of regulations, duplicative rules, and, in general, excessive costs.

Considerable controversy exists over the true impacts of environmental rules. Business sector allegations are apt to lack empirical backing and look only to the cost side of the equation, ignoring the societal benefits that derive from controls. The Conference Board report is not atypical; it has little convincing data and few examples of impacts. Environmentalists generally minimize the importance of cost, sometimes taking cavalier attitudes about effects on businesses and on the consumer (Frieden, 1981).

Part of the controversy results from the fact that observers of regulation are describing a number of phenomena. No one standard of performance is being regularly and consistently monitored. Complaints may come from businesses that overcomply in order to avoid any chance of citations for violations or from those who feel large companies use regulation offensively to price them out of otherwise competitive markets. Indeed all logical possibilities of compliance

---

[59] *Los Angeles Times,* Jan. 4, 1983, at 2, col. 6.

exist within American business; there may be undercompliance, overcompliance, and marginally acceptable compliance. The ways in which firms have organized themselves to produce any of these outcomes determine in part perceptions of the problem of noncompliance and the costs industry associates with environmental requirements.

Overcompliance is reported among companies who in order to insure against enforcement actions may add excess technology or overcontrol a manufacturing process. Automotive industry spokespeople report that emission controls add greatly to the cost of the final product because of the perceived need to build redundancy into the production system and because of the necessity to meet the most stringent conceivable standard that could prevail. To wait for promulgation of the actual standard before production and process design would not allow sufficient time to tool up to produce a complying product. This phenomenon is most prevalent when the industry is not successfully influencing the nature of standards.

The concept of overcompliance is similar to that of ''positive responsiveness'' offered in Roberts and Bluhm (1981). They use the term to describe ''taking the initiative and proposing new government rules and regulations'' (p. 4), which does not match our notion of overcompliance; but they point out that organizations can overreact to ''short-run fads,'' thereby doing ''more than would later seem appropriate once popular opinion had swung back the other way'' (p. 5).

The costs of overcompliance go beyond those reflected in conventional accounting. Those with initially strong environmental agenda can also ultimately suffer from overcompliance. Firms that are using resources wastefully on compliance are more likely to undertake both legal and extralegal attempts to avoid control than those that are satisfied with standards.

From a societal perspective, marginal compliance may be most efficient. Here industry uses just the amount of resources necessary to stay within the legal limits that government will enforce. Arriving at this optimal state is not easy; few firms report that they have achieved acceptable compliance at acceptable costs. Those companies that do manage just to meet standards do complain less about the aggregate expenses of compliance. As well, some commentators feel that this strategy is efficient only when cleanup costs are high; otherwise, from a societal point of view, exceeding legal requirements might be more efficient.[60]

The significance of available data also is a source of controversy. EPA reports that between January, 1971, and December, 1982, 154 plants were closed, in part because of environmental regulations. These closings affected 32,749 employees. Eighty-two closings involved failures to meet air standards;

---

[60] This point was made in a personal communication from John Braithwaite on reading an earlier draft of this book (letter of July 2, 1984).

54, failures to meet water standards; and the remainder, a combination of air and water or other regulatory focuses.[61] Whether this is an acceptable cost of regulations has no objective answer, although these numbers represent tiny fractions of the industrial base of the United States.

Industrial location decisions are also said to be related to the stringency of a state's environmental laws. From the environmentalist's perspective, statements that compliance will force relocation amount to blackmail; but for business they are no more than a prediction based on cost analyses. Again good data are difficult to come by, although some studies conclude that environmental regulations are not a significant factor in such decisions. In a 1983 analysis, corporate executives in explaining industrial moves ranked environmental rules behind market location, labor availability, access to raw materials, and quality of life (Stafford, 1983).

The statistical debate about the value of compliance will continue. Christiansen, Gollop, and Haveman (1980) may have summarized it most accurately: "Full appraisal of regulation must recognize unmeasured economic benefits not reflected in most analysis" (p. 2). Negative impacts of environmental regulation on industry have been selective; mining, particularly copper, and utilities and construction have been hardest hit. The authors conclude that regulatory schemes tend to be inefficient in that they focus on engineering-oriented, rather than performance, standards. This focus induces "a level of capital investment and capital intensity in excess of that required to efficiently achieve environmental goals" (p. 3). The effect of environmental laws on productivity growth rates was about one percent of the rate. The effect on employment was "not very severe." Overall, their thoughtful conclusions suggest why views on the benefits of compliance can be so polar.

> There is no real consensus on the relative magnitudes of the factors responsible for the slowdown in productivity growth. The slowdown in capital investment, the changing demographic composition of the labor force, the changing composition of output, and business cycle factors seem to be most prominent. The impact of changes in relative energy prices remains highly controversial. It seems clear that environmental and health/safety regulations bear some responsibility for poor productivity performance, but little evidence exists to suggest that as much as 15 percent of the slowdown can be attributed to them. A reasonable estimate would attribute from 8 to 12 percent of the slowdown to environmental regulations. This is not to say that all public regulations taken together (including those regarding, for example, new product introduction, transportation, plant location) have not had a major impact. In any case, difficult to measure economic benefits and various social welfare questions must be considered before a full evaluation of regulations can be made. (p. 5)

A major factor on the benefit side of the equation is the uncertainty of health

---

[61] EPA Economic Dislocation Warning System Quarterly Summary Report, 1982 American Statistical Index.

effects of noncomplying activities. Whether environmental pollutants are harmful enough to justify expensive compliance programs often is a question without a clear answer. Information necessary for unambiguous conclusions is unavailable. In many cases it will be unavailable for many years, if at all. Therefore value judgments must dominate determinations of the point at which regulation stops being rational and compliance represents an overly conservative risk assessment.

The desirability of requiring compliance also differs according to the individual regulation. A union may support a stringent nationwide environmental policy and yet side with management on a decision about controls on an individual plant. A farmer may value regulations that protect the watershed in which he works but consider occasional violations insignificant. Hazardous waste haulers may welcome national policies for their industry but oppose the costs of redundancy in a system which requires cradle-to-grave monitoring of all waste materials with the concomitant paperwork of a manifest system.

The Aims of This Volume

There is a strong need to study the effectiveness and efficiency of social legislation and the compliance systems associated with it. To waste resources on ill-conceived or poorly designed legislation makes little sense. . . . In order to design and maintain cost effective compliance systems, the underlying decision rationales that cause managers to comply or not to comply with such regulation must first be understood. (Greer & Downey, 1982, p. 488)

This book, then, aims to address the question: What policies will achieve the goal of persuading business to comply with resonable environmental law? This approach requires attention to the development of regulations and does not simply treat rules as inherently deserving of business obedience.

Some scholars consider the nature of the sanction to be of paramount importance in effecting business behavior in regard to goals expressed in law. Whether the sanction should be civil or criminal is a widely discussed issue (Clay, 1983; Glenn, 1973; Kovel, 1969; Morris, 1972; O'Brien, 1978). Whether the underpinnings of environmental law should be based on incentive or punishment is also often debated (Anderson, Kneese, Reed, Taylor, & Stevenson, 1977; Kagan & Scholz, 1981; Marin, 1979; Schelling, 1967, 1983).

Other factors can influence how business behaves. The way in which regulations are written and communicated can be important. So is the extent of support for compliance with rules by local or national interest groups. Attributes of government agencies, their professionalism, the kinds of resources on which they can call, the speed with which they choose and apply a sanction or incentive, the fairness of enforcement policy also bear upon outcomes.

Despite propositions about the impact of each of these factors, there is little convincing work regarding which matter most in influencing business. Packer (1968) bemoaned our ignorance of how criminal sanctions deter economic offenses; this ignorance generalizes to much of the analysis of the response of complex organizations to law.

This volume is based on the proposition that no single variable can explain compliant behavior. Factors interact to effect compliance differentially in varying business scenarios. The book presents a framework for understanding how this takes place.

Overview of the Work

The next chapter inventories approaches to compliance adopted by government or offered as improvements. The strengths and weaknesses of each are summarized. Options include criminal and civil sanctions, administrative strategies, cooperative or informal mechanisms, business volunteerism, and economic incentives.

We then introduce (Chapter 4) a framework for understanding compliance with environmental regulation which builds on systems theory and information exchange. Chapter 5 describes the behavior of enforcement. What assists or inhibits prosecutors in enforcing existing regulations? How important is the nature of the sanction? the size of the enforcement force? How important is the environmental law experience of the government lawyer as compared with that of business counsel? The significance of the clarity, consistency, rationality, and scientific basis of the law itself is then addressed (Chapter 6). Which characteristics of regulation influence business behavior? Chapter 7 investigates the roles which groups outside of government and business play in directing the enforcement of the law. Professionalism and status of the creators of the law and those who implement law are discussed. Does it matter whether environmental directives come from Congress or state legislatures, the courts, or administrative agencies?

Characteristics of business are investigated as factors that promote or inhibit compliance. Is there a criminogenic business type? Is noncompliance related to business size or profitability? How important is business involvement in development and interpretation of regulations?

In these chapters compliance is described as a process and the compliance event is analyzed. Compliance is seen as evolving from interaction among several groups, as occurring over time and as an outcome that is difficult to control by any single policy lever. Actors include environmental groups with specific societal objectives, businesses with profit and social responsibility motives, businessmen whose decisions are determined by diverse forces including personal advancement and the desire for association with one or other business ideology, experts within agencies who have a range of professional obligations and goals, and enforcers of environmental law who themselves have professional agenda ranging from meeting perceived quotas for citing violators to befriending industry officials who may be future employers.

These chapters identify no simple policies that might insure that small and large businesses will follow environmental rules. By itself, drafting regulations more carefully will not effect greater compliance, nor will the use of sanction x or penalty y or sentence z. Assisting environmental support groups will not automatically improve compliance in any linear way. And just as it cannot be demonstrated convincingly that capital punishment lowers the murder rate, policymakers should not conclude that any one factor, such as self-regulation, will increase business responsiveness. Each factor in the framework needs attention. The last chapter describes ways in which to improve the communication of regulations, to reform enforcement, and to strengthen support groups which promote compliance. Government and business can work to make compliance more acceptable—by having regulations more fully recognize business constraints, by increasing the knowledge that business has of environmental rules and their rationales, and by fostering business access to economical ways of reaching reasonable environmental goals.

A Word on the Study

A considerable body of knowledge exists on individual compliance, on the behavior of the firm, and on the nature of violations of regulatory law. Important studies come from the legal literature, from social psychology, from organizational studies, from economics, and from the interdisciplinary work on regulation. There is also an embryonic literature on the firm's response to regulatory law (Bardach & Kagan, 1982; Braithwaite, 1981, 1982a,b; Coffee, 1981; Di-Mento, 1976; Fisse, 1971, 1978, 1980; Geis, 1978; Hawkins, 1984; Kagan & Scholz, 1981; Note, 1976; Skolnick, 1966; Stone, 1975, 1978, 1980, 1981). In the first phase of this study these works and the expanding literature on regulation were reviewed to formulate a framework for analyzing business compliance. The framework then became the basis for data collection of several kinds. Case studies were undertaken. We interviewed executives and small businessmen about their firms' reactions to a charge of noncompliance or an enforcement effort and about their general attitudes toward securing compliance. We talked with corporate executives in their world headquarters and with small businessmen at their plants. We communicated with those jailed for noncomplying behavior. Our investigation took us through the compost piles of a mushroom farm, into fields where toxic wastes are handled and mishandled, and into the elegant suites of large corporations. Throughout the book findings from this part of the study are referred to as the *executive interviews*. We also conducted telephone interviews with present or former government attorneys in order to present a comprehensive summary of the cases presented in Chapter 1.

We conducted over sixty interviews of businesspeople, environmental and policy scientists, and government regulators. Interviewees included corporate vice-presidents, chief executive officers of small companies, senior environmental scientists in large firms, corporate counsel, division heads, and agency administrators and enforcement officers. Some sessions involved group interactions in which top managers exchanged views among themselves about environmental regulations; others were individual interviews. We hoped to promote candor and thoroughness in our interviews by a pledge of confidentiality and therefore we have not included a list of interviewees here.

We were able, for the most part, to talk frankly with targets of enforcement actions. Not all interviews were productive. A few interviewees gave only self-serving responses. On occasion, secretaries kept us from obtaining access to their employers. But most people reacted with curiosity, if not genuine interest, and welcomed the opportunity to comment on regulatory law and to describe their perspectives on individual cases.

Interviews were directed by an interview guide that contained both open-ended items and those that tapped specific factors in the compliance framework. Some respondents talked in broad generalities that addressed global issues in compliance. Others, including people in the laboratories, offered highly detailed information that can come only from daily work with the compliance challenge.

We assembled files on thirty other cases of noncompliance using secondary data. Data included legal pleadings, reports in scholarly journals, and media materials. A summary of the legal proceedings was developed from the time of the alleged violation until the most recent available indicator of response to negotiation or enforcement and of compliance and noncompliance. We coded the sanctions sought and the sanctions received, the strategy used in the choice of sanction, and the reported reactions of defendants and government and third parties. Our cases range from the nationally prominent and highly visible (*Love Canal–Hooker, Reserve Mining*) to those of only local interest (*A–Z Decasing, Capri Pumping*) A complete list of cases is found in the appendix.

We surveyed forty-four enforcement officers in state and federal agencies. A mailed questionnaire was returned by 66% of the sample ($N = 29$) drawn from jurisdictions differentially concerned with environmental enforcement and representing all regions of the country. Throughout the book, in reporting data, this group is referred to as the *surveyed enforcers*. Respondents were from state enforcement agencies (usually the attorney general's office), state environmental policy agencies, or federal environmental enforcement agencies. Respondents from thirteen state enforcement agencies represented twelve different states (Alaska, California, Delaware, Illinois, Maine, Massachusetts, Minnesota, Missouri, North Carolina, Texas, West Virginia, and Wyoming) and nine federal regions (1, 3–10). Respondents from twelve state environmental policy agencies represented nine different states (Alaska, California, Delaware, Minnesota, Texas, Illinois, Massachusetts, Vermont, and Missouri) and seven federal regions (1, 3, 5, 6, 7, 9, 10). Respondents from federal environmental enforcement agencies included two from EPA, one from DOJ, and one from DOI. Of the 29 total respondents, 27 were attorneys who had practiced for an average of 10 years (SD = 4.3 years, range = 3–22 years) with the majority (14) having practiced between 7 and 10 years. The 27 attorneys had been with their agencies an average of six years (SD = 3.7 years, range = 1–13 years) and the majority (15) had been with their present agency between 4 and 10 years.

Our questionnaire addressed attitudes on enforcing environmental law and on influencing business to meet environmental goals. It explored the factors that determine choice of cases, how government characterizes environmental violations, and views toward reforms.

We interviewed enforcement officers and inspectors of the South Coast Air Quality Management District; attended administrative hearings on petitions for extensions of compliance deadlines and variances and proceedings in California and federal district courts in several cases of alleged violation of environmental rules; and observed (and participated on) advisory committees for regional and federal environmental agencies.

This book is not an environmental law text. Knowledge of environmental law is not assumed.[62] When familiarity with such law is required for understanding of a point, the law is summarized.

---

[62] There are a few good texts in environmental law for those who desire greater substantive background in the field: Anderson, Mandelker, & Tarlock, 1984; Bonine & McGarity, 1984; Dolgin & Guilbert, 1974; Findley & Farber, 1981; Rodgers, 1977; Schoenbaum, 1982; Stewart & Krier, 1978.

CHAPTER 3

PURSUING COMPLIANCE

Society's Tools

The explanation of noncompliance is not that society lacks strategies for promoting environmental goals. Indeed, public policy's arsenal of sanctions and incentives is full. But the nature of the carrot or stick is not the most important factor in influencing business behavior. To understand this point it is first necessary to inventory available strategies. Choices theoretically range from draconian criminal sentences to subsidies for compliance. In reality, a variety of actors constrain choices in ways which we present in the following chapters. But it is useful to know the range of options for achieving compliance.

The legal and criminal justice literature has extensively evaluated sanctions (Ermann & Lundman, 1975; Geis & Meier, 1977; Packer, 1968; Stone, 1981). The most common response options are presented in Figure 1. When government is informed about a violation or a pattern of violations or when government obtains that information through inspection and monitoring, several choices are available. In some situations action is legislatively determined, as when an imminent threat to the public health is alleged.

The government can act administratively and issue a notice to comply; failure to do so leads to prosecution through judicial action, or in limited circumstances to direct restraining orders, as in the *Reserve Mining* case, or to administratively imposed monetary penalties. The agency can enter into a settlement agreement with the firm, negotiating an approach to compliance without calling on any other branch of government. Or it can seek injunctive relief directing the firm to act in a way consistent with law. In the alternative, the agency may seek civil or criminal penalties by referral to its own or a separate legal unit. In *Frezzo*

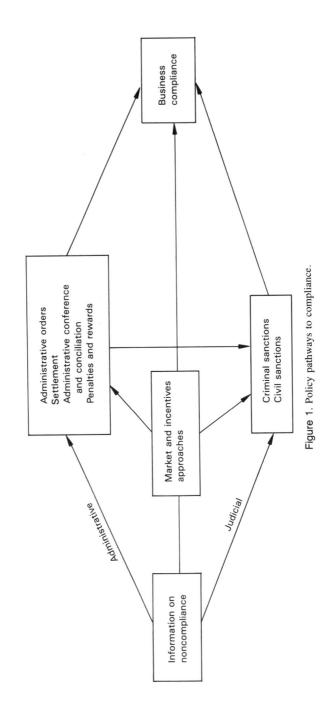

Figure 1. Policy pathways to compliance.

Brothers and *Distler,* discussed in Chapter 1, the government pursued this strategy.

Government also controls an immense amount of largess and determines property rights in society; some compliance strategies manipulate these powers. For example, the federal government has linked sewer and highway funding to compliance with requirements of the Clean Air Act. And government determines what sets of economic incentives are available to promote environmental quality.

Several factors influence the choice of strategy to secure compliance. The administration's regulatory philosophy plays a role. Of surveyed enforcers, 41% reported that "the political philosophy of . . . supervisors" is either somewhat or very important in decisions on cases to pursue in enforcement actions.[1] At one extreme is the view that violations of environmental regulations demand strict criminal sanctions. The opposite position imputes no responsibility to the firm; rather, compliance is achieved through polite interactions among reasonable people in government and industry. Such interactions may include subsidies from government to the firm, technology transfer, accelerated depreciation in a tax code for installation of pollution control devices, or direct grants to create prototypes of emission control systems.

The Criminal Sanction

Criminal sanctions are provided under both state and federal environmental statutes. For example, the Federal Water Pollution Control Act Amendments of 1972[2] authorize penalties of $10,000 or six months in jail for falsification of documents under the permit system and of up to $25,000 or one year in jail for willful violations. Government can also use federal or state criminal statutes involving mail fraud, perjury, obstruction of justice, criminal contempt, conspiracy, accessory after the fact, aiding and abetting, and racketeering. Other quite general legal provisions cover some violations. For example, in California, unintentional and inadvertent dumping of hazardous wastes has been successfully prosecuted under a general provision against "unlawful business practice."[3] And in Illinois the attorney general charged officials of Film Recovery Systems, Inc., with murder under an Illinois homicide statute. The victim allegedly was killed by exposure to hydrogen cyanide in the company's suburban Chicago plant.[4] Three company officials, the former president, the plant supervisor and

---

[1] Executive interviews discussed in Chap. 2, p. 36.

[2] 33 U.S.C. §§ 1251–1376.

[3] California Business and Professional Code §§ 17200 and 17203. One case, which also employed other counts, was *People v. The Garrett Corporation,* 3 C416199, Superior Court, County of Los Angeles. The theory was also used in a case against Southern California Edison for inadequate cleanup of PCBs subsequent to a rupture of a power pole–mounted transformer.

[4] *The Wall Street Journal,* Jan. 7, 1985 at 14, col. 1.

the plant foreman, were convicted of murder and the company itself was convicted of involuntary manslaughter.[5]

The Case for Use of Criminal Sanctions[6]

Advocates of use of the criminal law for environmental violations make several arguments. These center on its effectiveness as a deterrent and its appropriateness as a punishment. Rehabilitation of company personnel is also cited.

A primary rationale is that the criminal law raises the salience of government action. Unlike the civil sanction, the firm fears criminal law; through it government sends a message to the company that those in decision-making

WARNING
THE ILLEGAL DISPOSAL OF
TOXIC WASTES WILL
RESULT IN JAIL.
WE SHOULD KNOW
WE GOT CAUGHT!

Dear Businesses & Residents of the City & County of Los Angeles

Pollution of our environment has become a crisis.

Intentional clandestine acts of illegal disposal of hazardous waste, or "midnight dumping" are violent crimes against the community.

Over the past 2 years almost a dozen Chief Executive Officers of both large and small corporations have been sent to jail by the L.A. Toxic Waste Strike Force.

They have also been required to pay huge fines; pay for cleanups; speak in public about their misdeeds; and in some cases place ads publicizing their crime and punishment.

THE RISKS OF BEING CAUGHT ARE TOO HIGH—
AND THE CONSEQUENCES IF CAUGHT ARE NOT WORTH IT!

We are paying the price. *TODAY,* while you read this ad our President and Vice President are serving time in *JAIL* and we were forced to place this ad.

PLEASE TAKE THE LEGAL ALTERNATIVE AND PROTECT OUR ENVIRONMENT.

Figure 2. Adapted from an ad used by the Los Angeles City Attorney's Environmental Protection Section.

---

[5] *The New York Times,* June 15, 1985 sect. 1, at 1, col. 5.

[6] Throughout this chapter we offer specific arguments for each approach to compliance. Some criticisms apply to several approaches; they are summarized under the heading with which they are most frequently associated.

positions will hear. Peter Beeson, formerly with the United States Department of Justice, explained that prosecution may be more effective for crime in the suites than it is for crime in the streets: "Deterrence works best on people who have not had contact with criminal justice and for whom prosecution or even investigation will have severe personal consequences."[7]

Sanctions that have great publicity value are available under the criminal law.[8] Imprisonment of company officials is a possibility. Government can require community service of executives or mandate probation. Criminal fines can be imposed, and corporate personnel can be disqualified from serving on corporate boards. The public and the press pay more attention to these sanctions than to civil penalties. That a corporate executive may actually "serve time," that he or she may be made to perform a menial job, calls attention to the seriousness of the environmental violation.

In Los Angeles, a number of companies have agreed to publicize wrongdoings as part of a criminal sanction. For example, Precision Specialty Metals was required to purchase an advertisement in *The Wall Street Journal* describing its illegal activity and the effect of the violation on the public health. The advertisement was to include the fact that jail sentences were involved. The Los Angeles city attorney at the time, Ira Reiner, gave his rationale for this innovative sanction:

> We wanted the ad in *The Wall Street Journal* because that newspaper finds its way into every corporate boardroom in America. We want to put a chill in the boardrooms. We want them to understand that if they violate the law, they had best get away with it, because if they don't, they are going to jail.[9]

Fisse and Braithwaite[10] have chronicled how corporate executives worry about adverse publicity of law violations. The overall cost may be greater than paying a large fine. Adverse publicity can have impacts on sales and other long-term effects on the competitive posture of the company.

The failure of company officers to recognize the nature of environmental problems leads to certain regulatory violations. Requiring executives to undertake cleanup activity, to temporarily manage an environmental advocacy organization, or to engage in other community service can increase appreciation of environmental concerns. Fisse suggests that such community service orders prevent the corporation from delegating the task of compliance to "minions within

[7] Greider, "Fines Aren't Enough: Send Corporate Polluters to Jail," *Rolling Stone* 10 (March 29, 1984).

[8] A large percentage of surveyed enforcers (54%) reported that their agency seeks publicity on selective enforcement activity. Surveyed enforcers discussed in Chap. 2, p. 36.

[9] Greider, "Fines Aren't Enough: Send Corporate Polluters to Jail," *Rolling Stone* 10 (Mar. 29, 1984).

[10] Personal communications during preparation of Fisse and Braithwaite manuscript on corporate response to law (1982).

the organization. . . . Not to insure involvement by senior and middle management would be to allow the sanction to fall upon tools or scapegoats'' (1980, p. 12).

Some base their support of the criminal strategy on moral arguments. Imposition of imprisonment is said to remove the class bias in the control of deviant behavior, since the wealthy cannot avoid the pain of punishment through payment of insignificant fines. A version of *noblesse oblige* has also been articulated; the powerful have a greater responsibility to society: when they fail to meet their obligations, they should be subject to serious sanctions.[11]

Customized sentences are available in criminal sanctions. Experts in corporate motivation and corporate behavior can guide sentence choice. The National Center on Institutions and Alternatives is one group that establishes specific programs for felons. It acts in response to requests by lawyers and will specify the most appropriate type of community service, where the offender will reside, and approaches to supervision.[12]

Sanctions such as disqualification from holding corporate office can have systemic effects (McDermott, 1982). Government can impose true costs on a violator in the form of lost wages and lost opportunities since sanctions can preclude compensation or corporate indemnification of executives. Disqualification prevents future violations, at least in the roles which the executive held during the original violation, and disqualification may change the defendant's attitudes toward crime. McDermott writes:

> Sanctions which diminish the authority of lawbreaking actors increase the relative influence of lawabiding ones while simultaneously punishing lawbreakers by depriving them of wealth and other rewards derived from their exercise of power . . . In this vein, forced exclusion from office dramatically reduces the authority of the convicted executive in a milieu where company loyalty may on occasion be regarded as a higher obligation than obedience to the law. (p. 61)

McDermott concludes that the sanction may also force reevaluation of the company's attitudes toward violations.

Others argue that, even if not always effective, the criminal sanction must be used for some environmental violations. It is the only acceptable response to egregious antisocial behavior. Whatever the costs, society has no choice in certain cases but to prosecute with the full power of the law.

Describing a Los Angeles case in which the defendant allegedly dumped 4,000 gallons of highly toxic materials into county sewers through ''pirate pipes,'' the prosecuting attorney differentiated the actions from regulatory infractions:

---

[11] I am indebted to Gilbert Geis for these rationales.
[12] ''Defining New Terms,'' *American Way Magazine 32*, Nov. 1981.

> This is the most clearly willful act we have ever come across . . . We're not talking about some inadvertent disposal because of a technical violation of a complicated rule. . . . We're talking about real skulking-around-the corner stuff.[13]

Examples of such wanton disregard for the consequences of environmental violations are numerous.[14]

A controversial rationale for use of the criminal sanction in regulatory law is that it deters violations in ways superior to other approaches to achieving compliance (Kadish, 1963; Kagan & Scholz, 1981; Packer, 1968).[15] Deterrence is spoken of in two classes: special deterrence and general deterrence. In the special case, the criminal sanction is said to prevent the individual violator from committing subsequent violations. Surely this is true in the short run if the responsible executive is imprisoned. The more questionable effects are deterrence over time for the individual perpetrator of the crime and general deterrence, that is, the extent to which others are influenced to be more law-abiding by observations of the treatment of the criminal. The debate about deterrence of regulatory crimes is almost as long and heated as the argument over individual deterrence. Those who continue to advocate the criminal sanction for deterrence purposes must selectively choose within a rich literature (Blumstein, Cohen, & Nagin, 1978).

The Case against Use of Criminal Sanctions

Critics cite several arguments in opposition to use of the criminal sanction for getting compliance, or at least in opposition to its routine use. Interestingly, many of these are made by enforcement personnel themselves.

Some arguments center on procedure. Use of the criminal sanction calls into play the criminal justice system with all its safeguards for defendants, availability of jury trial, and evidentiary rules. The burden of proof is greater in a criminal case than in a civil case. In criminal law the state must prove its allegations beyond a reasonable doubt. In a civil case the standard of proof is the preponderance of the evidence. This difference can be extremely important in environmental law where securing good evidence and establishing causal links between a defendant's acts and actionable behavior are often difficult.

Locating actual violators within the firm is not easy; once located, attributing individual motivation and intent sufficient to meet criminal law standards is also problematic (Solomon & Nowak, 1980). If the state chooses to pursue the case by attributing responsibility to a high-level executive, other problems arise. Coffee (1981, p. 455) has noted:

[13] *Los Angeles Times,* Oct. 14, 1983, at 1, col. 5.
[14] *Distler; Derewel.*
[15] Only a small minority (10%) of the surveyed enforcers (discussed in Chap. 2, p. 36) noted the deterrent value of the criminal sanction, but those who did so described the effects unequivocally.

Under Model Penal Code Section 2.07(1) (C) , the prosecutor must prove that either a
"high managerial agent" or the board of directors performed, authorized, or reck-
lessly tolerated the events. This doubles the prosecutor's burden because in addition to
proving the crime occurred the prosecutor must also impute intent to high managerial
levels within the corporation; this in turn may create an incentive for high officials to
insulate themselves from such information.

In addition, under the criminal law opportunities abound for defendants to
forestall the impact of successful prosecution. Aspects of this response are totally
within the ground rules of the American legal system. Nonetheless, some procedural arguments exploit and manipulate safeguards designed to protect the
innocent—but thereby available to the shrewd guilty party as well.

Access to procedural challenges can also draw out prosecution in ways that
undermine its cost effectiveness and its potential to persuade. *Dow Chemical Co.
v. United States*[16] is a case in point. The EPA had employed aerial photography
to determine whether a chemical plant was violating air quality emissions standards. Dow had denied the EPA's request for entry to take ground-level photographs of the emissions from Dow's power houses. The United States District
Court found that warrantless air surveillance was a Fourth Amendment violation,
an unreasonable search, that Dow had a reasonable expectation of privacy that
the air surveillance had violated, and that this approach to monitoring exceeded
the EPA's authority under the Clean Air Act.[17]

Other analyses conclude that the cost–benefit ratio of using criminal sanctions is unfavorable. In the *Distler* case, costs incurred by public agencies for the
prosecution, including expert fees, were approximately $1.5 million. The cost
includes prosecution expenses, including a jury trial. The deterrent effect of
criminal sanctions in the area of environmental regulation is also disputed. By
the time an action is brought, those who should be affected by a criminal
information or indictment may no longer work in the firm. The expected value of
punishment may be small, since it is highly uncertain in areas of corporate or
white-collar crime. Coffee (1981, p. 455) has described a further problem in the
deterrence logic:

> Only low-level scapegoats may be disciplined by the firm, since passively responsible
> senior officials may be able to disguise their own involvement. The logic of deterrence
> may produce a highly biased form of internal discipline which never penetrates to
> upper levels.

Both business and environmentalists also cite political and social disadvantages of use of the criminal sanction. Reliance on the public prosecutor can be a
serious enforcement weakness. Prosecutors respond to incentive structures that

---

[16] 536 F. Supp. 1355 (1982).
[17] Under developing theories of privacy, the same constraint might also apply in civil proceedings. A
court of appeals overturned the district court opinion. As of this writing, it is not clear whether
Dow will appeal. *Dow Chemical Co. v. United States ex rel Burford* 14 ELR 20858 (6th Cir.,
Nov. 9, 1984).

do not necessarily track on promoting strong regulatory compliance records. Successful careers in the law will only rarely be made from pursuing pollution control criminals. Another industry perspective sees selective criminal enforcement as a way in which ambitious prosecutors attract attention, sometimes with little concern for the true merits of the case or the nature of the environmental harm.

Criminal sanctions may not be used even if available and appropriate.[18] On a mundane level, the local or state prosecutor may be ignorant of criminal enforcement capacity. Stewart and Krier (1978, p. 554) report that "no civil or criminal penalty was assessed for violation of environmental law in Connecticut in the period 1970–1974." In 1981, Representative Toby Moffett reported to the chairman of the Nuclear Regulatory Commission, Nunzio Palladino, that criminal penalties in the Atomic Energy Act which, in his words, "were not placed in the Atomic Energy Act as window dressing," were ignored as a means of achieving compliance. The commission was described as reluctant to act on criminal conduct, "particularly when the trail leads higher and higher in the utility management."[19]

Several studies of white-collar crime describe criminal sanctions as an empty threat (Geis & Meier, 1977): Neither juries nor judges will punish violators criminally. Only in the most egregious cases will the judicial system enact criminal sanction against the "upstanding" corporate executive.

Arguably useful in combating early notorious manifestations of a problem, the criminal sanction is deemphasized after it has addressed blatant wrongdoing. Freilich (1982) has concluded in analyzing housing code enforcement:

> The criminal sanction approach served as an effective measure of code enforcement until the 1950s and 1960s when the focus of housing code enforcement changed from eliminating serious, blatant public nuisance health problems to attempts at enforcing and supervising a multitude of violations as our housing standards increased. (p. 330)

Courts became unwilling to recognize housing violations as true crimes and practically never imposed jail sentences. These conditions are somewhat applicable to environmental law. Some violations, such as illegal disposal of toxic wastes, are linked to serious health problems. But much of the governmental response to environmental rule breaking has been routinized outside of the criminal justice system.

The Civil Sanction

The civil sanction is broadly available to promote environmental compliance. The sanction may be statutory or based in common law. Numerous

---

[18] Fifty-two percent of surveyed enforcers (discussed in Chap. 2, p. 36) reported that when criminal sanctions are available the enforcer does not pursue them.

[19] *Los Angeles Times*, 17 Dec. 1981, at 2, cols. 5–6.

theories of liability exist for environmental violations, and the law can expand these for problems particular to a historical period, such as those associated with hazardous wastes, acidification, smog, and noise. Government uses the common law through its enforcement arms and its environmental agencies directly; and nongovernmental entities, including individuals, employ common law theories through the courts. Two classic environmental cases are *Boomer v. Atlantic Cement*,[20] brought by a citizen's group against a cement factory under nuisance law, and *Spur v. Del Webb*[21] which involved a corporate residential home developer, a large feedlot operation, and purchasers of single family homes in a developing but still primarily agricultural area. In *Spur* the court reached a compromise among the parties relying on nuisance theory and the so-called "coming to the nuisance" defense. Plaintiffs also employ strict liability, trespass, and negligence theories, although less often than nuisance, to seek compliance by large corporations with general environmental principles (Anderson *et al.*, 1984).

Several states have adopted the Model Natural Resources and Environmental Protection Act. The act gives any legal entity, including a private citizen, the power to initiate a lawsuit against anyone "for the protection of the air, water and other natural resources of the state and the public trust therein." It goes beyond common law theories and is the epitome of the civil law's flexibility. The act removes all major obstacles to obtaining judicial review of allegedly noncomplying acts, including the threshold determinations of standing and exhaustion of administrative remedies.

The civil law encompasses several remedies. Fines can range from the nominal to the severe. An air pollution district may be limited to seeking monetary penalties for stationary sources (all nonmobile polluting entities) of one to two hundred dollars per violation; but under the Federal Clean Air Act[22] and Federal Water Pollution Control Act Amendments[23] civil fines of up to $25,000 per day are available for stationary and point sources and up to $10,000 per day for automobile emission control violations.[24] The Allied Chemical Company in a civil suit was fined $13.2 million for the Kepone spill; this was later reduced to $5 million. (The money was used to create an endowment for an environmental trust in the state of Virginia.)

Damages and injunctions are also available. Damages may be compensatory, those aimed at making the injured party whole, or exemplary or punitive, those aimed at communicating the desire to solace the damaged party or punish the violator. Damages of up to three times the value of the harm are available

[20] 26 N.Y. 2d 219, 257 N.E. 2d 871, 309 N.Y.S. 2d 312 (1970).
[21] 108 Ariz. 179, 494 P. 2d 700 (1972).
[22] 42 USC § 7413(b).
[23] 33 USC § 1319(d).
[24] 42 USC § 7524.

under antitrust laws applicable to environmental violations, but these have rarely been awarded.

Several state environmental statutes and federal antitrust law provide for the injunction (which is also available under criminal law). The injunction is a governmental decree requiring the termination of a violating activity or mandating an action to effect compliance. It can be temporary or permanent. Environmental law examples are plentiful. Under the Michigan Environmental Protection Act, utility condemnations, industrial air pollution activity, hotel expansion, and land drainage are among the many activities that have been enjoined.

The Case for Use of the Civil Sanction

Advocates of the civil sanction emphasize its flexibility. The civil law is widely perceived to be less costly and time-consuming than its criminal counterpart. Furthermore, many civil law theories, themselves applicable to a wide range of fact situations, are available under environmental laws and under tort causes of action to address a given violation. In addition, the law of civil procedure, while complex, is more favorable in certain aspects to plaintiffs. The standard of proof is more easily met, and fines can be allocated for uses directly related to compliance. Fines are revenue sources for environmental agencies. In 1981–82, 75% of the $21 million budget of the South Coast Air Quality Management District of California came from fees and fines. In addition, government can structure civil suits to meet the specific ends of environmental policy. A city attorney explained the main objective of a suit against a corporation which illegally dumped toxic wastes:

> The top priority is to guarantee a good water supply for that part of the county. The intent is not to cause a financial hardship on the corporation, and any civil penalties will be put into the abatement account. We are not trying to put them out of business or have people lose their jobs.[25]

The regulator can fashion civil remedies such as the injunction to address the specific causes of noncompliance. Government can enjoin a firm from using its sewer outlet to dispose of toxics; each such use is a violation. A manufacturer may be instructed to produce an emission control device according to detailed standards. Government can direct a handler of hazardous waste to indicate the nature of the contents of drums and to report those contents according to published schedules.

Finally, civil actions impair the business posture of the firm. They can require that any pending civil suit against the company be reported to stock-

---

[25] *Sacramento Bee*, Dec. 27, 1979, at A22, col. 3, quoting Los Angeles Deputy Attorney General Joel Moscowitz.

holders. Fear of the investor's response is an additional incentive for compliance. Treble damage awards can raise the monetary threat of the law to preclude the firm from simply treating the fine as a cost of doing business. In addition, the antitrust law fosters competition that promotes compliance (Goldstein & Howard, 1980). The promise of royalty payments to the exclusive owner of a control technology acts as an incentive to push the state of the art and deters collusion to limit best available technology. The antitrust approach forced the large automakers to compete to develop emission control equipment that would set the industry standard. And the threat of civil sanctions will impede industry from keeping information necessary for standard setting away from responsible federal agencies; those pressures also diffuse responsibility for the development of controls.

The Case against Use of Civil Sanctions

Criticism of use of civil law to promote compliance comes as general attacks on the capacity of the civil law to alter organizational behavior and complaints against individual civil law approaches.

A fundamental alleged weakness is that civil law does not communicate sufficient concern about the "evil" of environmental violations. Society must treat noncompliance with more than organizational wrist slapping. The Connecticut chief state attorney in requesting funds for an additional criminal attorney and investigator put the matter this way: "The civil means is not going to do it when people are flouting the criminal law on a daily basis."[26]

Specific civil remedies are also criticized. Potential defendants in environmental law cases focus on the injunction, a civil remedy, found in some broad environmental laws such as citizen suit legislation. Industry argues that these statutes do not mandate sufficiently clear standards to allow a defendant to determine what is complaint behavior. The injunction gives judges excessive authority to formulate requirements without technical expertise and tempts plaintiffs to harass industrial targets with idiosyncratic notions of what is an acceptable pollution control record. Some government agencies have also opposed bills which would introduce the injunctive power[27] because citizen suits can disrupt long-range planning for environmental quality by administrative agencies. Regulators cite two kinds of problems. Suits pull agency personnel who have to prepare for trial and to testify from important planning tasks. And citizens challenge and enjoin innovative programs before their merits are tested. Citizen civil suits also allow costly and unnecessary challenges to environmental standards that are formulated through participatory procedures.

Moreover, while large fines are theoretically available, the average amount

[26] *The New York Times*, Feb. 18, 1983, at B8, col. 6.
[27] See Sax and DiMento, 1974; and DiMento, 1976, 1977.

of a civil penalty is not sufficient either to promote general deterrence or to truly influence an individual violator. Recent successful resolutions of air pollution cases in California are illustrative. In one reporting period the total fine that industries paid for forty-five violations, including court costs, was less than $15,000. The largest fine, for excessive emissions, was $800. The large fines reported in the popular press are newsworthy because they are uncommon.

Courts are reluctant to impose sizable civil fines because of the perceived negative economic effects on employers, employees, and stockholders. And some courts view fining the company as counterproductive to the compliance goal: they deprive the violating firm of a potential source of funds for developing or installing control technology.

Finally, those who have been subject to it attack the antitrust sanction and its associated treble damages provision. Critics cite behavior opposite to "technology forcing" or promoting innovations in pollution control equipment. In the patent cross-listing case involving the major automobile manufacturers, Chrysler argued that the competitive incentives supposedly contained in the Clayton Act[28] upon which the suit was partially based were illusory.

> The competitive instincts, which the antitrust laws encourage, are the precise opposite of those which conduce to improvement of the environment. . . .[29]

Chrysler continued in a brief:

> In the absence of collusion, an individual automobile manufacturer would be extremely reluctant to install such a [pollution control] device. Unlike, say, a stereo radio, the device would add to the cost of the automobile without improving the product from the standpoint of the purchaser, since he would benefit only trivially from the reduction in pollution brought about by the device. For this reason, the manufacturer could not recoup any significant part of the cost of the antipollution device in a higher price for the automobile unless his competitors also installed such devices. If, therefore, manufacturers are permitted to agree on the introduction of antipollution devices, the devices will probably be installed faster.[30]

Administrative Sanctions

Government can employ administrative orders to promote compliance. Statutes usually write in this agency authority. Powers include revoking a permit or license, imposing administrative fines, or directing specific complying acts. A well-known example is the EPA's order which empowers the agency to require auto manufacturers to call back automobiles that exceed air quality standards.

[28] 15 U.S.C. §§ 12–27 (1976). The antitrust law is also subject to analysis under the criminal sanction because criminal penalties are also available. See, for example, 15 U.S.C. § 2.
[29] Quoted in R. B. Stewart and J. Krier, *Environmental Law and Policy: Readings, Materials and Notes* (Indianapolis: Bobbs-Merrill Co., 1978), at 298.
[30] *Id.*

The manufacturer is required to repair the cars without charge to the customer.[31] If not contested by the manufacturer, the approach requires no judicial procedures. Another example is the fund cutoff for regions that do not meet pollution control standards. Other supplementary enforcement power may be lodged in administrative agencies. When an imminent and substantial endangerment to human health or the environment exists, the agency can immediately proceed to close a violating operation. An example is the approach in section 6973 of the Resource Conservation and Recovery Act which provides: "The Administrator may also, after notice to the affected state, take other action . . . including, but not limited to, issuing such orders as may be necessary to protect public health and the environment."[32]

In the well-known Times Beach dioxin case, the Missouri Hazardous Waste Commission acted administratively to halt transport of hazardous wastes by the company which had earlier sprayed dioxin-tainted oil as a dust control measure. At the very least, administrative orders can slow down activities that are considered detrimental to environmental quality.

A novel and controversial use of the administrative order is the non-compliance penalty[33] adopted by the EPA and earlier used in Connecticut. Under this power the agency calculates the amount of money that a violator saves by failing to come into compliance with environmental regulations. Once calculated, this figure is charged to the firm, almost like a bill for services.

Arguments for and against Use of Administrative Orders

The debate over withholding federal funds for regional noncompliance highlights one form of the administrative order. The government is empowered to terminate certain federal funds if an air quality region fails to meet ambient standards or fails to make adequate plans to comply by legislatively dictated deadlines. These monies include the much sought after highway and sewer construction grants and the five-cent-per-gallon gasoline tax. Grants involved are those for projects which would increase environmental degradation of the region.[34]

---

[31] See the description of the process in the *Chrysler* case in Chap. 1. A federal court of appeals limited the reach of this power to newer cars with less than 50,000 miles of use. *The Wall Street Journal*, Dec. 19, 1983, at A–7, col. 1. But the court later reversed itself. See *General Motors*, the Appendix.

[32] 42 USC §§ 6973.

[33] The Connecticut Enforcement Project, 1973, Con. Pub. Act No. 665. See the Clean Air Act Amendments of 1971, 42 U.S.C. § 7413 (D)(3). The formula under the Connecticut Enforcement Project was that the agency may assess a firm up to the amount of the full benefit of not installing pollution equipment. The agency assessment has the same legal effect as a judicial judgment when docketed in a court. See also text accompanying fn. 11, Chap. 5.

[34] An example is limitation of funding assistance for Jackson County, Oregon, "because the state of Oregon failed to submit and is not making reasonable efforts to submit a legally enforceable State Implementation Plan (SIP) revision . . . for carbon monoxide." 50 FR 8116, Mar. 4, 1985.

The approach recognizes one factor that drives business behavior: the economic incentive. But the wisdom of using the fund cutoff as a compliance strategy is questionable. Fund cutoffs can prevent implementation of projects which are developed to meet goals under other governmental policies. Among those that this sanction would affect was a garbage-burning plant in Brooklyn, New York. The plant was designed to utilize recycling techniques that the federal government has been supporting. A less dramatic example of counterproductive results is termination of air quality planning, an activity also threatened by fund cutoffs. Federal funds for inventorying emission sources and devising strategies for reaching NAAQS are withheld, thus preventing a district from doing its work because of prior difficulties in reaching standards.[35]

Practical obstacles to the use of the strategy also exist. Formal rule making under the Administrative Procedures Act[36] is required preliminary to the cutoffs. These procedures take time and mandate participation by affected parties. Companies invariably cite economic disruption and other adverse impacts related to loss of federal funds; these arguments make the sanction politically sensitive.

The fund cutoff may be a paper tiger. In 1982 industry expressed little concern over a threat to employ the sanction widely in regions that had not met NAAQS by the 1983 deadline. Business concluded that it was inconceivable that the EPA would actually "close down the country."[37] Environmentalists also opposed the proposed action, fearing that fund cutoffs would have a backlash effect: the Clean Air Act would appear so unreasonable that Congress would move to amend it.

There appear to be ways of circumventing construction fund cutoffs once put into effect. A widespread ban on federally funded development in southern California's six largest metropolitan areas reportedly had no effect on industrial growth "because the EPA [found] ways to allow permits to be granted for large projects that otherwise would have been stopped."[38]

A major weakness is that the administrative order is often challenged. Industry perceives the "letter from the administrator" as a noncompelling first step in protracted interaction with government. If not satisfied, the firm can challenge the order in court. A corollary weakness is that the agency feels pressure to devise administrative orders that will not be contested; this tendency often translates to erring on the side of leniency.

Conditions necessary for successfully implementing effective administrative orders are not common. One condition, the firm's predisposition to

---

[35] Exemptions from funding limitations are provided by regulations. See 45 FR 53382 Aug. 11, 1980. These include "projects which improve treatment capability but do not expand capacity for future growth."

[36] 5 USC §§ 551–559; 701–706; 3105; 3344; 5262, 7521, (1967).

[37] *Inside EPA*, Dec. 3, 1982, at 9.

[38] *Los Angeles Times,* Feb. 1, 1983, at 1, cols. 1–2.

comply, limits the range of violations the order will influence. The EPA concluded in 1981 that it would employ RCRA orders only in situations

> where there are reasons to anticipate a high likelihood of compliance by the potential respondent. Factors to be considered in evaluating the likelihood of compliance include indications of respondent's good faith and readiness to address the problems, apparent ability and disposition to commit the resources to accomplish the necessary remedial action, and the complexity of the cleanup measure required.[39]

For companies with limited resources the weakness of the administrative order is quite the contrary; the administrative order does not insure the procedural due process protections available in civil or criminal proceedings. Business fears arbitrary use of the order and abuses of administrative authority.

A main strength of the administrative order is its simplicity. The processes of effecting compliance remain with the environmental agency and the violating firm; the judiciary need not be involved; the adversarial nature of getting compliance is minimized and cooperation between government and industry is fostered.

In practice the agency can use the orders selectively, simultaneously countering any argument that an economic sector will be ravaged and putting that sector on notice. The potential of this sanction is considerable. Use of administrative orders can be inexpensive and flexible. When employed professionally, they help build direct communication links with business. Orders can also substantively reflect the regulator's direct knowledge of the firm in ways not often found in judicial outcomes.

Less Formal Approaches and Their Strengths and Weaknesses

Depending on one's view, environmental law is either plagued or blessed with a number of less formal approaches to fostering compliance. Some of these are steps before resort to criminal or civil sanctions; others are substitutes for sanctions—means which themselves are sufficient to achieve reasonable environmental goals.

Conference and Conciliation

Conference and conciliation are the oldest members of this class. Administrative agency representatives consult with potential and actual violators and discuss ways of improving performance. By becoming aware of business difficulties in reaching environmental standards, government both better understands the firm's actions and suggests more effective means for achieving compliance.

---

[39] From EPA Enforcement Guidance, Sep. 11, Douglas MacMillan, Acting Director of the Office of Waste Program Enforcement, quoted in *Inside EPA*, Sep. 18, 1981, at 8–9.

Furthermore, business will be more receptive to environmental objectives if government communicates the nature of goals and their rationale.

Critics argue that without specific schedules for coming into compliance and absent milestones of acceptable behavior, business has little incentive to alter polluting behavior. Proponents counter that conference and conciliation are cost-effective when compared to command and control and concomitant litigation; the positive business–government relationship that evolves is an important precursor of ongoing compliance. Legal commentary tends to be skeptical, but enforcement officials consider informal strategies an important part of programs designed to induce compliance. Only 14% of surveyed enforcers[40] described informal procedures including conference and conciliation as "generally ineffective"; 28% concluded that they were generally effective; and 55% reported that the effectiveness of the procedures varied.

Numerous other individual strategies for promoting compliance exist. The government can publicize lists of wrongdoers. Sometimes referred to as "black listing," the approach combines the government publicity power and the power to withhold largess. The regulator publishes names of those who have failed to meet clean air, clean water, or other environmental standards in newspapers of general circulation and in industry oriented media. Government may follow up and give companies on these lists low priority in disbursement of contracts and grants, which total $200 billion per year.

The regulator can authorize use of the *environmental audit* (Braithwaite, 1982). Several variations of the audit exist. A firm may voluntarily establish an in-house auditing team or use a third party contractor to monitor compliance with environmental standards. The company then suggests an auditing structure which government approves (or modifies and then approves). Once the blueprint is accepted, companies are required to report violations and to describe activities to bring the firm into compliance.

A variant is a totally voluntary audit. The private sector sets industry guidelines for reaching pollution control goals which government establishes at only a general level. Industry also determines the timetable for arriving at milestones. The sources themselves are responsible for monitoring compliance. The audit strategy is based on a belief that, for the most part, noncompliance is an aberration that does not result from any deliberate business attempt to circumvent environmental rules.

Proponents argue that the environmental audit lessens the adversarial relationship between business and government. By employing audits, the regulator avoids the wasteful pursuit of penalties when business is making good faith attempts to meet pollution control goals. The audit is flexible and government can combine it with policies such as expedited permit processing for companies

---

[40] Survey of enforcers discussed in Chap. 2, p. 36.

with good compliance records. Also, the self-directed audit may be more effective in identifying truly serious violations: industry knows that those in and out of government who have acceded to its argument that it can police itself will judge it strictly.

Other support for the voluntary strategy comes from recognition of the practical problems of effective enforcement in a highly regulated society. Resources are not sufficient for systematically pursuing violators under command and control models. In the environmental arena literally billions of pollution sources exist; even in major industries, government inspections take place only once or twice over a period of years.

Some theory in law and society favors the self-audit and self-regulation over command and control strategies. Teubner (1984) says the latter are based on simplistic notions of linear causality ("the goals determine the program, the program determines the norm, the norm determines changes of behavior, these changes determine the desired effects") and bloated understandings of the information that the regulator has about the regulatee. In contrast, self-audit and self-regulatory strategies rely only on reasonably available information; they efficiently discriminate serious violators from occasional noncompliers; and they avoid excessive costs of imposing stock approaches on very different firms (Braithwaite, 1982; Scholz, 1984; Teubner, 1984).

The weaknesses of volunteerism are patent. The American Bar Association Committee on Environmental Control enumerated some. Data supplied by the industry on its own performance may be of uneven quality, complicating both assessments of the extent of actual noncompliance and the regulator's priority setting. The potential for bribery of the auditor exists. Voluntary approaches may require the creation of another bureaucracy, the function of which is to collect and process compliance information.

Business itself does not unanimously support the audit strategy. Some companies fear that government will use information gathered in the audit to initiate enforcement actions. The recent promotion by public interest lawyers of a new variety of "citizen enforcement" will make the audit even more worrisome for industry. In this strategy, citizens are trained to read compliance records and encouraged to review them in federal and state environmental agencies.[41]

Environmentalists argue that the audit is akin to "letting the fox guard the chicken coop." An industry that is expected to regulate itself must resolve the conflict between self-interest and the interest of the consumer. Industry, which has always had the option of self-regulation, usually resolves these issues at the consumer's expense.

Critics say voluntary means are incapable of influencing the behavior of

---

[41] Greidel, "Fines Aren't Enough: Send Corporate Polluters to Jail," *Rolling Stone* 10, (March 29, 1984).

complex organizations. Volunteerism will be misused and abused; sophisticated firms will manipulate cooperative strategies—appearing to behave responsibly while failing to meet control goals.

Negotiation-Based Approaches

A special class of less formal approaches to effecting compliance involves negotiation and mediation of environmental issues and controversies. To achieve specific compliance, parties may negotiate disputes, and as a strategy for promoting general compliance an increasing literature advocates negotiation of rule making.

Negotiated strategies can proceed in parallel with other efforts to get compliance. Representatives of parties to an issue, dispute, or policy set the agenda for exploring and resolving differences. At times they may call upon the assistance of a third-party mediator but they need not do so (Sullivan, 1984). As a device for settling disputes, the short track record of negotiations is quite strong (Sullivan, 1984) and the embryonic use of negotiation in rule making has also been promising (Russell, 1985). Sullivan reports successful negotiation in a controversy between the Montana Power Company and the Northern Cheyenne Indians over construction of two 700 megawatt coal-burning plants; in an agreement to construct a dam on the North Fork of the Snoqualmie River in Washington; in siting and design of a shopping center in White Flint, Maryland; and on conversion of an oil-burning power plant to coal in Brayton Point, Massachusetts.

The advantages and disadvantages of negotiation are speculative because of the short experience with the innovation. The EPA employed an early form of negotiated rule making in the development of a regulation involving application of the secondary treatment requirement under the Clean Water Act, and in 1980 Senator Carl Levin of Michigan introduced a bill which would have provided for regulatory negotiation in the federal bureaucracy. It was only in 1982 that the Administrative Conference of the United States recommended that agencies undertake negotiation of proposed rules (Russell, 1985). Russell reports consensus among numerous stakeholders in two prototype regulatory negotiation efforts at the EPA. Subsequently, EPA, OSHA and the Food and Drug Administration initiated negotiations. One involved the proposed nonconformance penalty rule; the other, an emergency exemption provision under the Federal Insecticide, Fungicide, and Rodenticide Act (FIFRA). Both negotiations led to rules which were sent to the EPA administrator to be published in the Federal Register as Advanced Notices of Proposed Rulemaking.

Proponents argue the case for negotiation to achieve specific and general compliance citing both behavioral and technical advantages. Bacow and Wheeler (1984) conclude that negotiation, as contrasted with some other forms of rule making, involves parties who have to live with regulatory outcomes and who are able to articulate their preference directly. These authors say that negotiations act as a model for expeditious subsequent problem resolution among parties. Sullivan (1984) asserts that negotiation puts a premium on cooperation. He contrasts consensual approaches with command and control strategies:

> With final decision-making resting in the regulatory agency or court and issues circumscribed by technical concerns, each side in a dispute faced incentives to portray the adversary's goals in the worst possible light and frame their objections in legal terms. (p. 14)

Others claim that negotiation exposes the public to the challenges of fashioning good rules and that it promotes mutual respect among the negotiators: "This is, perhaps, the single most lasting outcome of these [negotiated rulemaking] demonstrations" (McGlennon, 1985, p. 497).

Technical advantages of negotiation include increased probability that environmental issues will be addressed substantively (Bacow & Wheeler, 1984) and that issues will be analyzed fully from multiple perspectives as opposed to the narrowing strategy of command and control (Sullivan, 1985). Negotiations can reduce decision costs and delays in comparison with the protracted process in more formal approaches (Susskind, Bacow, & Wheeler, 1983), and they promote education of all parties about contingencies and trade-offs in regulations (Susskind *et al.,* 1983) and about the burden, especially to industry, of compliance (Bacow & Wheeler, 1984).

Several weaknesses of negotiation remain. Some problems are practical; some, legal; and others involve more perplexing behavioral uncertainties. Sullivan (1984) notes that it is difficult to determine who is to participate in negotiation and that front-end costs of the process may be considerable. He also warns that disputing parties could use direct communications "to intensify their animosities" (p. 183).

Bacow and Wheeler (1984) warn of the conceptual difficulties of resolution of complex technical issues; perhaps they cannot be negotiated. Put another way, achieving consensus in negotiation does not necessarily lead to compliance. Implementation must follow. This is so for any approach to getting compliance, but the focus in consensus-based approaches may make implementation less central than is the case in other forums.

Furthermore, leaders may not be able to bind members of the organizations represented in negotiations to the outcomes of consensual proceedings (Bacow & Wheeler, 1984)—a problem less severe in traditional judicial proceedings because of doctrines which preclude litigation on issues already resolved. Other strictly legal objections: under the Administrative Procedures Act, rule making must be done on the basis of a record the creation of which is incompatible with certain versions of negotiation; some models of negotiation may involve illegal *ex parte* communications (which are oral or written communications with the agency that are not included in the formal record); and negotiated rule making may make it difficult for an agency to describe the basis and purpose of the rule (Note, *Harvard Law Review*, 1981). Wald (1985) notes that judicial review of outcomes of negotiation

> opens up its own Pandora's box; it is likely, indeed, that the bifurcated standard and the reliance on consensus will continue to entangle courts in the regulatory process. To determine whether negotiated rules are within an agency's "jurisdiction," for example, courts will still have to ensure that the rule falls within an agency's statutory perimeters and does not conflict with a specific congressional interest—a none too easy task . . . if the rule is challenged, the courts will be left with the task of determining which interest groups must be represented and what, if any, type of participation by other individuals or groups is appropriate. (p. 526–527)

Although some commentators describe the legal challenges as major, means for overcoming them include making negotiations public; having the agency articulate standards that limit the negotiators' discretion; and offering procedural safeguards to those who would be represented in rule making. Furthermore, advocates usually offer the reform as a supplement to lawmaking procedures, not as a substitute.

Incentive-Based Approaches

Economic incentives are alternatives to more traditional means of inducing compliance. Although the approaches differ, all assume that the firm behaves in an economically rational way and that government can structure policy initiatives to make compliance economically attractive. Some sanctions within the legal system also address business's economic concerns, such as criminal and civil monetary penalties. The approaches differ in that a legal wrong is not at issue.

Environmental policy has access to a large inventory of economic incentives. *Effluent charges* are fees which may be administratively imposed for discharge of pollutants into the air or water or onto the land. Under the concept of the *marketable permit*, government issues a predetermined number of "licenses" to emit pollutants. These permits can be allocated by market mechanisms and be awarded to the highest bidder, or regulators can distribute them in other ways, such as by rationing according to targeted classes (manufacturers and preservationists, for example). The EPA has applied the concept to the control of chlorofluorocarbons.

Control trading refers to programs that allow firms to make decentralized intrafirm or interfirm decisions about the least expensive way of reaching standards. The bubble concept is the most well known; it employs the fiction that an imaginary bubble can be superimposed on any pollution source. Since the concern of those seeking compliance is with the total effluent from that bubble, how industry controls effluents is a matter of indifference. The firm itself can choose the production process to alter or the emissions to restrict on the basis of economic and other considerations. The concept can apply to water and air pollution control, within and across plants, and even across industries. Proponents describe regional, national, and even global bubbles.

Offsetting or trading and purchasing of rights to pollute is a variation of this idea. Emissions sources decide among themselves how to achieve a standard. If an industry creates an increment of pollution, the regulator will require it to purchase control equipment or to pay another source to purchase that equipment to remove an equal or larger increment. The approach was contemplated when SOHIO sought permission to establish a major port in the city of Long Beach, California, to transport oil from Alaska to the East Coast. Under a proposed but never implemented plan, SOHIO would provide funds to small polluters, including dry cleaners, in southern California for installation of additional equipment. The plant selected dry cleaners because they are, in the aggregate, a significant source of air pollution in the Los Angeles basin; yet they are notoriously poor compliers. The owners of these small companies claim to be ignorant of regulations and to lack sufficient funds to install effective pollution control equipment. The SOHIO plan would have allowed large industry to finance emissions control activities outside of their own boundaries or bubbles—on sources where available technology works best.

Government compliance strategies may rely on direct economic incentives. *Tax policy* is a pollution control lever. Government can provide tax relief to those who install control technology by creating favorable amortization periods for certified air or water pollution control facilities and by offering investment credits and accelerated depreciation.[42]

Probably the most extreme application of economic incentives is the direct payment to companies which have histories of pollution. Since the industry that pollutes also creates societal benefits such as employment opportunities, some argue that government should directly subsidize any activity that is necessary to limit effluents.[43]

The Case for Use of Economic Incentives

Advocacy of use of economic incentives centers on notions of fiscal and administrative efficiency. Proponents say that reliance on market mechanisms promotes least-cost ways of reaching control objectives both for the firm and for government.

These analyses employ several assumptions from classical economics. In-

[42] See Internal Revenue Code § 169.
[43] A number of other incentive ideas has been offered, but those listed are the group most often discussed.

centive approaches make the greatest use of data about both pollution sources and costs of cleanup. Required information is decentralized, that is, the polluting sources themselves possess it. Industry knows best the nature of environmental problems and the ways to control effluents, and it will make less expensive choices than centralized control agencies might (Anderson *et al.*, 1977; Drayton, 1980).

Economic incentives are said to advance technology with concomitant advantages to compliance over the long run. Since incentives make it economically rational for the firm to limit its effluent, industry will secure the best possible technology or process for doing so. Self-interest acts to control to the level of least cost—not simply to control to a standard set by government. For its part, government need not specify technology; it simply sets overall goals and leaves to the marketplace the means for reaching them. Administrative efficiencies will follow.

Advocates say market approaches reduce the adversarial nature of securing compliance. A small bureaucracy will routinely monitor performance and bill firms for their emissions. Government–business interactions will routinize and government will no longer have to attend to wrongdoing; it becomes an accountant. Industry itself decides its level of performance, calculating noncompliance as a cost of doing business.

Of the surveyed enforcers[44] 55% reported that in general they favored use of economic incentives. As a tool for promoting compliance, 42% favored use of the bubble concept (14% opposed); 24% favored use of marketable permits to discharge pollutants (44% opposed); 41% favored effluent or emission charges (34% opposed); and 41% favored direct government subsidies to purchase pollution control technology (31% opposed).

Flexibility characterizes the incentive approach. For example, these strategies need not completely replace command and control regulation; rather, they can supplement standards. Government would require all pollution sources to install technology or alter processes so that a baseline of emissions is achieved. Above this baseline, the company could choose to add controls or pay per unit of pollution emitted.

The Case against Use of Economic Incentives

Adoption of economic incentive approaches has been limited and slow. The environmental community, industry, and government have argued well against the approach.

A fundamental point is that adopting pricing mechanisms is not a trivial

---

[44] Surveyed enforcers discussed in Chap. 2, p. 36.

matter. Take the emission charge. It is not obvious how to set initial fees. The regulator could set fees high at the onset and allow a change downward if fees excessively inhibit industrial production. But maintaining production is also a societal aim and political opposition may make high charges difficult to implement. Furthermore, the choice of fee structure can vary with the objective of the charge system. EPA staff members have pointed out that different schemes will be associated with goals of overall efficiency, compliance with standards, promotion of cost-efficient controls, or reward of good faith attempts in the absence of fully compliant behavior.[45]

Ironically, industry opposes economic incentives in part because of the uncertainty associated with them and in part because of the potentially greater predictability of results of this strategy. Costs across government administrations and even within administrations may be quite variable, making corporate planning difficult. (What will be the fee for disposing of an effluent into a stream or into the air two to three years from now?) And where there is fee stability, industry is not so well able to discount costs than in the case of the almost random and yet uncommon imposition of fines under command and control regimes.[46]

Incentive schemes may also be unfair to certain sectors of industry. Braithwaite (1981–82), citing Rose-Ackerman, notes that because effluent taxes are set at levels that are optimal in the aggregate but not necessarily in the individual case, they may force firms to close. This is true "even when they could have remained in business if they were required to pay only the social costs imposed by the untreated portion of their waste" (p. 497). Wolozin (1968) predicts discrimination against the smaller firm as a result of difficulty in monitoring larger companies:

> The reasoning here is simple, and it hinges on the cost of abatement equipment. The smaller the firm, relative to the average firm in industry, the greater the probable financial burden of installing air pollution control installations. In one industry, it has been reported, the cost of an air pollution control installation would exceed the value of the production facilities of many small firms in the industry. (p. 237)

Several commentators make a more fundamental criticism of economic incentives. Industry does not follow a classical profit-maximizing strategy. How decisions on investment are made is not well known; but students of the business firm understand with increasing clarity that "economic man" is too narrow a conception to allow for good predictions of organizational behavior. Strategies based on manipulating economic variables are at best organizational experiments.

---

[45] *Inside EPA*, July 24, 1981, at 4.

[46] Certainty can impede government business cooperation because little room remains for explanation and recognition of extenuating circumstances under some economic incentive approaches: Said one enforcer survey respondent, incentive strategies "can lead to compliance with resentment and hard feelings."

Critics raise a variety of other questions. Will the existence of incentives in any medium, say air, promote illegal disposal in other medium, say land, which is subject to less certain scrutiny? Also, government monitoring of effluents may be difficult and bureaucracies may grow up not dissimilar in size and complexity from those within a regulatory framework. Stewart and Krier (1978, p. 519) ask simply but provocatively ''Does the Internal Revenue Code provide an example of what a fee system might become?''

Emission charges and market mechanisms raise legal issues. One involves abuse of the tax power. Is the charge a fee, penalty, or tax? If it is a tax, is the legislature using the tax power to bootstrap itself into arenas where it cannot constitutionally regulate? If a tax, is the legislature applying it in an (illegal) nonuniform manner? If a tax, how can it apply to promote municipal and state environmental compliance, since these governments may be immune from federal taxation? Other legal challenges are that fees are a denial of due process of law (is property being taken for public use without adequate procedural safeguards?) and equal protection (does the incentive strategy discriminate unfairly against certain targets of compliance?). Furthermore, if the particular incentive is characterized as a monetary penalty plan, it may involve an improper delegation of legislative authority to an agency. Some state courts may invalidate such action under a questionably appropriate but still vital doctrine that prohibits certain delegations of legislative authority. If the charges legislation ''impedes the 'flow of commerce' in national markets, or it and preexisting federal legislation pursue mutually exclusive solutions to the same environmental problem'' (Anderson *et al.,* 1977, p. 128), it may be invalidated under the interstate commerce or preemption doctrines respectively. Anderson and his colleagues document these legal challenges and offer reasonable approaches to surmounting most of them; nonetheless, the legal uncertainty surrounding some incentives strategies remains an obstacle to their adoption.

Regulators cannot morally or legally apply incentive innovations to certain forms of pollution. The strategies are not acceptable for control of toxic effluents, no matter how much a firm is willing to pay to use common areas for disposal. At one time this constraint appeared to be narrow. But scientists now conclude that the nature of toxicity is not sufficiently understood to allow confident labeling of some substances as nontoxic. Furthermore, a pollutant may be environmentally benign as it leaves an industrial source, but its combinations with other chemicals and materials may be ecologically damaging. Put another way, the synergistic impacts of pollutant A and pollutant B may be greater than their additive effects. Therefore, incentive programs that treat pollutants individually may inadvertently lead to the creation of new forms of environmental noxiants.

The strategy is also subtle and difficult to comprehend. Whereas many economists find incentive-based approaches theoretically elegant and sensible,

the economics profession has not been very convincing to industry and policymakers. The South Coast Air Quality Management District has discussed for
years emissions charges to supplement the command and control program, but
many questions remain: Is the strategy a license to pollute? Would source requests be treated as variances or is the emissions charge notion independent of
the traditional permit process? What is entailed in the concept of "full social
cost"—the term used to describe the amount to be charged for emissions?

Critics object philosophically to incentives regardless of their potential efficiency. The approaches are a "right to pollute" which government should not
tolerate. (The point has also been made that, if revenues from market strategies
are allowed to go into the general fund, government may conclude that allowing
pollution is a way to balance budgets.)[47] Similar ideological opposition characterizes evaluation of direct payments to industry. Both ethical concerns and
worries about the potential spillover of the concept to other areas of regulation
are voiced. Will we next give industry the option to install additional occupational safety devices or to pay into a special worker compensation fund? Should
airline safety decisions be left to the market? We can fly cheaply if we avoid the
costs of additional protective devices and aircraft guidance, fire protection, and
automatic landing equipment (Huber, 1984).

Finally, opponents say some incentives policies subsidize only the capital
costs of controlling pollution whereas a significant part of the noncompliance
problem results from improper and inadequate operations and maintenance.
Krier (1971) comments that tax policy favors physical improvements over possibly more efficient process changes. Direct payments can have similar effects.
The federally funded sewer construction program is an example: It influenced
even small rural communities to construct large centralized systems which can
have poor compliance records if professional maintenance workers are not readily available.

Using Society's Tools

The regulatory agency, either on its own or in cooperation with business,
has access to a great variety of tools for bringing about compliance. Connecticut
pioneered in use of economic incentives. Vermont has relied on the tax power to
curb environmentally destructive development. Pennsylvania and New Jersey
have been leaders in challenging environmental violators with jail sentences.
Michigan has made considerable use of the civil injunction. The federal government has alternated, both across and within administrations, in relying on civil
penalties, administrative conference and conciliation, and criminal law. The last

---

[47] This view was offered in the enforcer survey, discussed in Chap. 2, p. 36.

months of the Carter presidency heard a loud call for criminal penalties, but these sanctions dropped out of favor in the transition and during the early years of the Reagan administration, only to be resurrected by the end of the first term. Lately, both the federal and state governments have begun experimenting with a new generation of cooperation in the form of negotiation. Finally, regional environmental enforcement units, such as air quality management districts, usually display almost all strategies to business targets.

What drives the government's choice? The regulator appears to select from its enforcement catalogue on the basis of shifting societal conclusions about what is necessary to control environmental crime or promote protection rather than on any careful look back at what has generated specific and general compliance. Philosophies of government, short-term political agendas, and appreciation of a cycle of the seriousness of regulatory violations—rather than political science—most often direct what is pulled from the regulator's tool bag. An environmental disaster calls for criminal penalties; an economic slowdown provides the grist for advocacy of cooperation. And bureaucartic reticence and standard operating procedures explain why some strategies are known but not used. As for some tools for compliance—negotiation and economic incentives among them—regulators want to see successful application by others before they add a regulatory innovation to their repertoire.

Theoretically, government can employ the full panoply of the criminal justice system in targeting an individual firm or a whole industry. The regulator can sit down with industry and design programs for compliance or leave that design to the private sector. Economic incentives can be the basis of achieving compliance: these vary from directly paying firms which are not meeting environmental goals to imposing effluent fees invariably. Government can court, coerce and cajole.

What should the regulator do to promote compliance? Which tools should it use most often?

This book takes these questions and reframes them into a larger question: What system of government–business interaction most effectively promotes compliance? In the next several chapters we build the argument that compliance is a process that comes from exchanges among the regulator, the regulated industry, and other groups in the population who care about environmental quality.

Our answer to the question of efficacy of tools for compelling compliance is based on a review of the work on alternative sanctions. Strategies are differentially effective in differing circumstances. And our answer comes from development of a framework that explains compliance only in part by the sanction or the incentive involved. About the relative efficacy of the tools we inventoried only rather gross conclusions are merited. And even most of these are negative conclusions about the power of policy tools. Civil fines will not influence large syndicated violators of environmental law. Large established companies will aggressively counter the use of criminal penalties. Companies will accept administrative controls if they are not draconian and not publicized. Nonreputable industries will ignore the positive incentive to compliance incorporated in the tax code.

Even these generalities will be contingent on other important factors. Government's ability to focus influence on a business target is the essential factor in explaining compliance. The form of influence will vary. The means to focus influence may be a criminal penalty in one case and an invitation to jointly write better environmental rules in another. A strong enforcement policy is only one factor in our theory; the nature of the sanction is critical only in a very narrow band of cases. The tool which combines with other means of regulatory influence attends to the business motivation to comply. The means to manipulate that motivation may be a civil or criminal penalty or it may be a reward.

No one sanction is efficacious in the diverse factual situations which cause environmental violations. Threat of a jail term for corporate executives may be the only reasonable response to one class of violations. It may lead another business target to undertake effective measures to counter government policy. For a particular small company, precise use of civil penalties may affect the cost benefit calculation for responding to law just enough to promote compliance. A larger firm may be unmoved by the threat and, in developing a compliance profile, look rather to the regulator's recognition of its unique operating demands.

We shall see in the next several chapters that compliance takes several paths. The regulator must perform a fairly sophisticated analysis of the context of rule violations to determine which of its tools to call upon. In some circumstances, it will be most effective to make that determination in consort with business.

CHAPTER 4

THE BEHAVIOR OF COMPLIANCE

A Short Introduction

The environmental decade has been described as long on symbolic response to pollution and short on effective controls. A concomitant view is that environmental laws are so legislated and so administered that meaningful compliance is rare. Statistics on noncompliance (Chapters 1 and 2) and arguments against strategies for getting compliance (Chapter 3) generate negative views of the efficacy of environmental law.

However, other conclusions can be reached. In some industries compliance is more the norm than the exception. Violations make headlines because of their rarity. Furthermore, when violations are found or suggested, means exist to limit them. In this chapter we introduce a perspective on how compliance is achieved.

We present a theoretical framework (in Figure 3) that calls on systems thinking to describe compliance. We focus on communication of regulations; on enforcement of regulatory policy; and on characteristics of government regulators, of business firms which are targets of environmental law, and of groups which take a special interest in environmental quality.

The Framework Introduced: The Compliance System

The framework comes from several disciplines. We have reviewed and integrated research findings on individual, group, and organizational compliance in sociology, law, psychology, criminology, economics, political science, and management. The literature is vast and rich. It addresses compliance with law and with public policy directives. But it neither converges upon nor is integrated to address industry's compliance with regulation. Our task, therefore, has been

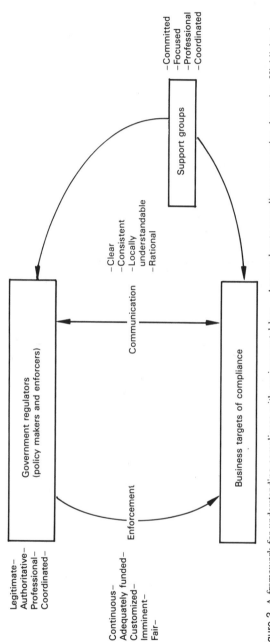

Figure 3. A framework for understanding compliance with environmental law. Arrows denote compliance promoting dynamics. Highlighted are compliance-promoting characteristics of actors.

to create a compliance framework. As Roberts and Bluhm (1981) note in their study of the power industry:

> We run the risk that our results will seem incomplete, unrigorous, and almost certainly unorthodox to practitioners in all of the disciplines. This risk seems worth taking, however, if it enables us to provide better ways of explaining, predicting, and controlling organizational behavior, which is the goal that provoked our research in the first place. (p. 9)

The framework is dynamic. At any one point, it captures, like a photograph, how actors relate in the compliance event, but the relative impact of the factors shifts over time explaining shifts in compliance. Actors who determine compliance can learn from their regulatory experiences. The firm monitors government activities to assess how clear, consistent, and rational communication of rules is. A company evaluates continuity of government policy and enforcement changes linked to administrative shifts. Rather than simply receive rules, a company will try, in ways presented more fully in Chapter 6, to avoid regulatory surprises. Similarly, government agencies can learn which strategies work best with various categories of organizations. Support groups have also learned, in two decades of environmental activism, how to affect compliance positively.

The framework incorporates both open systems theory (Cyert & March, 1963; Katz & Kahn, 1969) emphasizing organization–environment interactions (Lawrence & Lorsch, 1969) and closed systems thinking (Lippit, Watson, & Westley, 1958). The enterprise may occasionally act like a closed system, sometimes to its disadvantage and sometimes to its protective advantage. To survive, however, the firm must often behave like an open system, learning from its environment and being molded by the environment while attempting to mold it.

Organizations differ in their permeability to legal inputs. As in piercing the corporate veil, the regulator will sometimes be successful and sometimes fail to influence company administrators. Similarly, the business firm will, with varying success, influence environmental policy.

Government, business, and interested outside groups also try to develop protective barriers. Support groups try to keep secret their litigation and other priorities; government does not wish to telegraph enforcement strategies; firms have information on noncompliance that they hope will not be seen outside the company.

In the last two decades, companies have found regulatory demands ever changing and difficult to predict. The regulatory process itself can shift, including the procedures which guide the institutions that generate substantive rules. The business environment has been turbulent (Emery & Trist, 1965; Michael, 1973).

Systems theorists have developed a large inventory of the inputs and outputs of organizational life. (Allison, 1971; Cyert & March, 1963; DiMento 1976; Emery & Trist, 1965; LaPorte, 1971; Lawrence & Lorsch, 1969; Lippit, et al.,

1958; Mazmanian & Sabatier, 1981; Michael, 1973; Nagel, 1974; Note, 1976; Pennings, 1973; Roberts & Bluhm, 1981; Sabatier, 1977; Wilensky, 1967). Relevant exchanges between the firm and its environment are numerous. We have used two strategies to identify significant inputs and outputs (or exchange materials). In our review of the literature we have found a great emphasis on information. In addition, our case studies and the enforcer survey offered direction. Respondents told us what was most important in deciding how to comply or how to get others to comply, that is, what inputs they identified and what outputs they cycled. The two strategies for identifying important interactions are linked since doubtless we selectively heard what our theoretical choices had highlighted.

Of most significance are exchanges—not of raw materials or natural resources—but of information. Information can be policy, scientific data, propaganda, political influence, or legal pressure. We conceptualize the cycling of materials as primarily how information moves within and across organizational boundaries. We see its influence as directly correlated with certain dimensions of its quality.

In Figure 3, arrows represent several messages: the regulations themselves, enforcement messages, and information about the nature of violations of the firm and about the characteristics of the regulated firms or of the regulated industry. Obstacles to the transfer of information can take several forms and may be temporal: termination of communication may be the primary problem.

Even clear and consistent messages may be obstructed within an organization (Coffee, 1981; Stone, 1975). For example, channels within the firm or the regulatory agency may misdirect information. Coding processes may enhance or diminish reception. Support groups within the firm may direct regulatory communications to the proper place within the company or heighten the significance of information. And organizational complexity may distort, misdirect, or lose regulations.

Compliance can result from several different paths. Support groups alone may influence a firm to comply. Enforcement strategies, even in the absence of good regulations, may foster compliance. Or combinations of forces may explain compliance: when the regulator seeks and obtains good information about the firm; when enforcement messages are clear and consistent; when support groups within the firm and public interest groups converge on an understanding of acceptable behavior. Conversely, several different weaknesses in relationships among the government, the business sector, and environmental advocates can explain noncompliance.

Although single factors can be influential, only in the narrowest band of cases will one psychological, organizational, or economic factor explain compliance. We might say that systems attain compliance through a variety of routes. Individual dynamics may be critical, or combinations of factors, even when each

is not fully effective, may move the company to comply. Two examples anticipate what we will develop in Chapters 5 to 7:

1. Clearly stated environmental regulations based on respected scientific
   research may promote compliance even when enforcement policy is not
   fully implemented and when the general population is indifferent to
   compliance.
2. A strong enforcement policy which is fairly and predictably applied and
   backed by considerable resources may promote compliance even when
   environmental rules are inconsistent and lack scientific certitude.

As these examples hint, public policy should consider the route to compliance and not simply its realization since some routes are costly and
undesirable.

The Framework Elaborated

Chapters 5 to 7 individually address enforcement, rule communication, and
characteristics of those who participate in the compliance event. Here we introduce our theoretical framework in a summary form.

Enforcement

We focus first on information about the sanction employed in enforcement.
The effective regulator will base enforcement choices on considerable
knowledge of how the firm operates. The sanction, whatever its kind, must affect
the firm's complex cost–benefit calculations (Marlow, 1982); for, to summarize
roughly, business complies when all the information available on the meaning
and importance of rules and on enforcement indicates that costs of noncompliance will exceed benefits of noncompliance. The firm does not evaluate
its compliance decisions only in monetary terms; complex organizations have a
variety of corporate goals (Stigler, 1970; Wilson, 1980) emphasizing at various
times status, goodwill, and reputation.

Since perceptions of costs and benefits associated with compliance may
change over time, so may enforcement strategy. The firm may get used to
regulatory messages as consistent regulations help establish a kind of baseline of
the cost of doing business (Roberts & Bluhm, 1981). To be effective the regulatory agency needs access to information about the firm's subjective calculations of costs and benefits as well as the company's competitive profile (Diver,
1980). Such information, although not readily available to the regulator, would
be part of an ideal enforcement effort.

To affect the firm's decisions, the regulator must occasionally be prepared to impose high costs. In practice, several factors obstruct government's willingness to do so, and strong penalties are rare (Fisse, 1980; Stone, 1981). Neighborhood support of the challenged business activities and more widespread political support for the industry are two factors that explain such reluctance.

The literature on sanctions underscores a regulator's dilemma. For although the sanction must threaten high costs to the firm, if a large percentage of the target group considers penalties excessive, government's ability to foster general compliance will be jeopardized. Businesses will choose to fight sanctions considered unfair. They can litigate, use political influence to try to limit agency activities, and challenge the rationality of rules with propaganda campaigns.

Business is aware of other fairness aspects, including choice of targets. If business perceives that government chooses targets in a systematic and equitable way, both general and special compliance are fostered. But when government prosecutes only vulnerable firms in order to meet enforcement quotas, business may oppose enforcement policy. Small companies without resources may be vulnerable in one period, and uncommonly profitable companies may become clear targets in another.

The business community must see the sanctions as imminent. Communication of the enforcement policy has a time dimension (Chambliss, 1966; Ermann & Lundman, 1975; Gibbs, 1968; Horai & Tedeschi, 1969; Rodgers, 1973). Complex organizations can discount the costs of enforcement by projecting them into some uncertain future.

Naturally, the perception of imminence is linked to the regulator's ability to gather information. Institutional barriers to the flow of information between business and government can weaken enforcement threats. Certain agency decision rules formally impede information gathering. These include rules that promote societal objectives other than compliance—such as protection of proprietary information and privacy. Some obstacles are inherent in all conflict resolution procedures, but business can create unique barriers to environmental compliance (Marcus, 1980). Information considered irrelevant in other areas of law may also counter the threat of a sanction. For example, in determining penalties for environmental violations, courts consider the status of white-collar defendants and how punishment may affect the defendants' families. The business sector is aware of this history of sympathetic judicial treatment.

Information overload counters the regulator's ability to apply sanctions quickly. Business promotes overload in legislative, administrative, and judicial activities. Industry can supply immense amounts of data, some irrelevant, which the regulator must formally and systematically process. This need to react to information, no matter what its quality, impedes enforcement.

Something else, quite mundane, influences information flow. Many actors within the compliance event perform in ways characteristic of a bureaucracy faced with rapidly changing regulations. Therefore not out of malice, but rather because of inability to process information efficiently, government enforcement proceedings are often long, laborious, and inefficient. This state of affairs reflects the unrealistic design of some compliance objectives rather than substandard organizational performance.

Whatever the sanction, it must be communicated continuously (Ericson & Gibbs, 1975; Schwartz & Orleans, 1967; Skolnick, 1968). Otherwise enforcement messages lack credibility. Several organizational factors determine whether the message is continuous or aborted, as do relationships between government and the business sector, government and outside interest groups, and industry and support groups.

We emphasize the characteristics of the communication of sanctions. But what about the nature of the sanction itself and the relative efficacy of criminal, civil, and administrative sanctions? How effective are informal procedures and incentives to comply?

The nature of the sanction itself is less important than other factors in compliance. No enforcement strategy, when considered alone, universally motivates the corporation to behave.

The criminal sanction can deter, and studies have shown that its selective use is essential to deterrence (Kagan & Scholz, 1981). Its counter productive outcomes are also evident. Chapter 3 summarizes its strengths and weaknesses. Criminal sanctions can jeopardize cooperation between business and industry and raise the cost of violations to the point at which business chooses to oppose regulations. Criminal sanctions may chill a legitimate and useful challenging of environmental rules. They can counter a business–government learning process about effective regulations and freeze the cost of compliance at an excessive level.

Civil and administrative sanctions also have their weaknesses (Chapter 3). Whatever the sanction, government must use it reflecting knowledge of what makes the firm "tick." For its part, if business has information about the development of enforcement policy and if policy is intelligent and fair, industry is less likely to oppose enforcement activities actively.

Our compliance framework emphasizes information about enforcement. However, information may also counter compliance, and information overload is one of the agency's major problems. Furthermore, too much information about enforcement strategies may allow predictions of the probability of prosecution. Business may comply for a short period and then ignore regulations.

Communicating the Law

Characteristics of communication of regulations affect compliance. Environmental regulations must be understood by business; they should be neither overly specific nor vague. In short, they must communicate to the business targets in order to promote compliance (Coffee, 1981; Diver, 1983; Dolbeare & Hammond, 1971; Krislov, 1972; Lowi, 1969; Rodgers, 1973). The business target must know that a regulation exists before it can consciously comply.

A good deal of theory instructs that a mutual process of learning about the form of regulations improves compliance (Coombs, 1980). Public policy can promote learning in several ways. Participatory forms of rule making can enhance knowledge about both the constraints on the firm and incentives that motivate industry compliance. Rule-making strategies such as negotiation can customize information used in regulation. However, policy must guide participation and other learning approaches to avoid an imbalance between the government and business. Business can overly influence (or ''capture'') policymaking, or government can overwhelm an industry and jeopardize its capacity to compete.

The regulatory message must be consistently and constantly communicated (Krislov, 1972; Pfeffer, Salancik & Leblebici, 1976; Schwartz & Orleans, 1967; Skolnick, 1968). Such ongoing articulation acts as a reminder and an expression of commitment to implement law and contributes to a stable business environment. Constancy of message also has its limits; if government indicates that it will occasionally temper enforcement with flexibility, business respect for regulation will be greater than when messages are constant but unnecessarily rigid.

Several obstacles to information channeling jeopardize continuity in rule communication. These are found in agencies which articulate rules, in regulated industry, and in units which monitor or enforce rules. Excessive organizational differentiation in both business and government, absence of coordination within bureaucracies, and disruption caused by litigation are among the factors which impede the flow of regulatory messages.

Compliance is promoted if the regulatee considers regulations rational (Bardach & Kagan, 1982; Levin, 1977; Rodgers, 1973; Sabatier, 1977; Skolnick, 1968). Rationality has both objective and subjective dimensions. Whatever its form, it conveys to business that government cares about the quality of rules. Scientifically sound regulations also educate the firm about the environmental and health effects of pollution.

By focusing on this scientific basis, government can also forestall legal attacks which bog down the regulatory and compliance processes. Information, scientific and otherwise, can become weaponry in regulatory battles. Both government and business use evaluations of the quality of information in attempts to enlist support in the larger population.

The regulated firm evaluates rule rationality in both substantive and strategic ways. Business challenges rules because it directly shoulders the burdens of cleanup under tight schedules when environmental problems are not well understood or well described. Business will distinguish policies which are based on sound scientific background from those which transcend science and are based on value positions. Industry sometimes considers the latter more vulnerable to litigation.

Such challenges comprise a giant obstacle to gaining compliance. The nature of scientific inquiry is partially responsible. The structure of science invites attacks on methods and results, and the scientific process is fundamentally one of rejecting hypotheses rather than asserting causation. The same characteristics of science which promote sincere charges of irrationality also provide business with opportunities for tactical strikes at the information base of environmental law.

Our framework also holds that if rules are numerous, targets will be predisposed to question their rationality. The business presumption is that the information base for each rule must be sacrificed to volume of rules. The presumption does not follow logically, but in most regulatory settings it is merited.

The Actors

People are involved in the processes that create compliance and noncompliance. Our framework focuses on traits of public agencies, of business itself, and of those groups which care about environmental costs and benefits.

The firm's information about the regulator influences compliance. The business target evaluates the status of the government agency and uses this information as a barometer of the regulatory threat. Within government, individual administrators and small working groups contribute to a strong or weak reputation. The tenure and professionalism of administrators also affect how a firm assesses the need to comply.

Agency structure and organization are important to the delivery of regulatory information. Since the quality and strength of the regulation and its enforcement greatly influence compliance, the capacity to generate effective rules is crucial. Agency overload, inadequate resources, limited access to business information, and technical deficiences obstruct the ability to create, disseminate, and enforce good rules. Examples of poor regulation affect the way business views the regulator in general. Conversely, when government solicits and incorporates sound information into its rule making and effectively communicates environmental law, business takes notice. Business antiregulatory propaganda campaigns are less potent.

Aggressive labeling of actors promotes miscommunication in the compliance event. When business views environmental agencies in adversarial terms, or government uses ideology to categorize regulated industries, obstacles to compliance result. By heightening the conflict, both sides exaggerate the problems of incorrectly hearing the other. This is not to say that criticism is unfounded. Incompetence in critical business and regulator roles does exacerbate miscommunication and leads to attempts to force compliance using poor information. Neither command and control nor cooperative approaches are immune from these difficulties.

Another important actor is the support group. Interest groups of all kinds are concerned with environmental law. Some aim to secure strict compliance. Others emphasize societal goals which they see as inconsistent with strong environmental law. In our framework important groups are both small *ad hoc* gatherings which focus on a single firm or issue and larger, well established interest groups which seek or oppose strong environmental policies.

Information is a major source of support group influence. Support groups describe to regulators the impacts of environmental violations. They help establish priorities in a context wherein comprehensive enforcement is seldom possible. They alert business to citizen concerns about compliance. Conversely, support groups tell government when enforcement is unwarranted or inappropriate in light of the benefits of an enterprise.

The quality of information that support groups send varies. Inaccurate information can complicate routine and standard agency approaches to inducing compliance. Fixating on a particular case or procedure can impede development of new and effective compliance strategies.

Institutional channels modulate the strength of the compliance message. For example, citizen suits may make concerns of individuals or small groups heard when they might be ignored in administrative or legislative proceedings. Support groups can join together to enhance the import of information and to provide combinations of statements which are more influential than their parts. These coalitions also carry a symbolic message about the seriousness of environmental violations.

Just as information flows within environmental agencies affect the compliance outcome, so too does movement of regulatory messages (about rules themselves and enforcement threats) within the firm. The impact of regulation is mitigated by legal and interpretative resources available to the firm prior to rule implementation, and businesses vary in their capacity to influence regulations and to moderate their economic impact. Yet even some well-endowed firms may find it difficult to adjust to regulation. Organizational size and complexity may get in the way of simple and inexpensive compliance by obstructing information or postponing its communication to a control point. Corporate culture also acts as a filter of rules. Simple messages may take on exaggerated meaning or become a *cause célèbre* within firm A, while firm B economically complies or successfully interprets away the rules.

This summary introduces the compliance framework. The next three chapters develop each of the factors. In Chapter 8 we recommend ways in which information can be improved and channeled to promote compliance. In some circumstances better communication of rules is essential to reform; in others we call for improvement in enforcement and increases in support group influence.

CHAPTER 5

"THEY TREATED ME LIKE A CRIMINAL"

Sanctions, Enforcement Characteristics, and Compliance

During our trial . . . our attorney asked to bring in a cross-section of the community which we were affiliated with, as character references.

We could have brought two busloads. We brought in approximately 50. They were . . . firemen, policemen, other growers and just personal friends. This impressed the judge in the case so much he reduced the fines and sentence considerably . . . just on this.[1]

I am writing to express my astonishment and indignation over the Board of Public Works' withdrawal of a $105,000 fine imposed on Culligan Deionized Water Service, Inc.

I cannot believe that such a fine would "put the company out of business." Indeed if The Times *article (July 22) is accurate Culligan accepted between $5 million and $25 million from its clients before dumping the untreated waste into the city sewage system (with virtually no overhead but the cost of running a truck to the nearest sewer). If anything, the fine should have been much higher.*

I also cannot believe that the remaining penalty—jail for the Culligan president—will effectively deter such violations in the future. Serving 90 days in a minimum-security country club with other white-collar criminals seems a small price to pay for endangering the health of sanitation workers and the general public (especially when the profits of such crime continue to draw interest in the bank).

A punishment that better fits the crime might be to sentence [the executive] to spend 90 days personally cleaning up the sewers that his company contaminated.

—L. M. Dryden
San Gabriel[2]

---

[1] Interview with James Frezzo, December 28, 1981.
[2] Letter to the Editor, *Los Angeles Times,* July 29, 1983, Metro Section, at 4, col. 4.

Enforcement: Necessary But Not Sufficient

Efforts to achieve compliance with environmental law most often focus on a strong enforcement policy. As in many other areas of the criminal justice system, the almost unavoidable conclusion is that laws do not function smoothly because of a weak or nonexistent enforcement approach. Not enough violators are identified; when identified, not enough are sanctioned; and when they are sanctioned, penalties are insufficiently severe to communicate the fact that violations will not be tolerated.

This perception, as it relates to environmental law, is partially accurate. In the broadest understanding of the term, a sound enforcement policy is necessary to affect corporate and small-business behavior. But enforcement is to be understood to mean a system: A full enforcement system encompasses the sanction, the resources of the enforcing agent, the severity and certainty of a punishment's being imposed or an incentive's being awarded, the manner in which the regulated business perceives the enforcement policy, and the enforcement agency's relationship with other branches of government.

The Nature of the Sanction

Which sanction works best to promote compliance with environmental law? The question has no unambiguous answer. Chapter 3 introduced several perspectives on the efficacy of alternative sanctions. Here we delve deeper in order to survey the views of scholars, business people, and enforcers on the calculations undertaken in evaluating threats.

In a famous discussion of the criminal law, Packer (1968, p. 354) characterizes as "totally unexplored" the question of what makes the criminal sanction effective when "respectable" offenders are involved: conviction, jail sentences (imposed or actually served), publicity. "Our information about this kind of issue is mere impression and anecdote." Part of the reason for lack of knowledge has been that in criminal cases conviction rates for regulatory violations have been low and imprisonment quite rare.

Although some theorists hold that the deterrent effect (and ultimately the compliance rate) is greater for criminal law than for civil sanctions (Glen, 1973), others feel that in the areas of social and economic regulation, simply making an activity criminal is not sufficient to effect deterrence (Stone, 1978). Still others feel that, practically, there is little difference between criminal and civil law since the public does not equate moral turpitude with corporate environmental wrongdoing.

We can learn something about the efficacy of criminal sanctions in environmental law from empirical studies of deterrence. But the answers are limited. A body of deterrence research suggests that severity of punishment deters criminal behavior (Bean & Cushing, 1971; Erlich, 1973; Gibbs, 1968; Gray & Martin, 1969; Tittle, 1969; Tittle & Logan, 1973); but some researchers reach opposite conclusions (Schwartz, 1968; Waldo & Chiricos, 1972). Not only are research results contradictory, but application to the organizational target is not direct. Only in a few cases have empirical results involved the corporate criminal (Braithwaite & Geis, 1982).

Nevertheless, some theory quite strongly posits a deterrent effect of punishment in the organizational setting (Braithwaite & Geis, 1982; Chambliss, 1966; Packer, 1968). Chambliss considers white-collar criminals as eminently deterrable because their actions are often based on calculated risks, not on passion, as is the case for many other criminal behaviors. Braithwaite and Geis (1982) conclude that the reorientations which deterrence based on punishment require are more easily made by organizations than by individuals: "A new internal compliance group can be put in place much more readily than can a new super ego" (p. 310). Braithwaite and Geis (1982) summarize:

> The evidence on the deterrent effects of sanctions against corporate crime is not . . . voluminous. But the consensus among scholars is overwhelmingly optimistic concerning general deterrence. This may in part reflect an uncritical acceptance of the empirically untested assumption that because corporate crime is notably rational economic activity, it must be more subject to general deterrence. (p. 304)

But the effects of punishment must be better understood before public policy concludes that criminal sanctions are the enforcement mode of preference, for the empirical work on punishment leaves open two important questions for regulatory compliance: Might effects related to punishment derive as well from noncriminal ways of influencing business? And even if criminal punishments are effective, might not other incentives be equally or more effective with fewer costs and negative or regressive outcomes? Before concluding that public policy should rely mainly on punishment to gain compliance, further analyses of decision making within the firm are necessary.

The Business Perspective

In general, business people consider criminal sanctions for environmental violations, although acceptable in the abstract, infrequently appropriate and rarely applicable to their situations. This finding from the executive interviews is no surprise, but the passion with which business people attack the use of criminal sanctions may be (see also Clay, 1983). Business contrasts occasional "inadvertent" or "negligent" noncompliant activity with serious health and safety problems.

Clay (1983) reports that although a majority (54%) of interviewed California business people concluded that jail terms should sometimes be imposed upon violators of occupational health laws, most concluded that criminal sanctions were not appropriate in the cases in which they were charged with California OSHA violations. Respondents (90%) felt that the sanction should be reserved for serious violations committed by others in industry (Clay, 1983). Some enforcement officers concur. That criminal enforcement "tends to involve only fringe elements of the regulated community" was an attitude expressed by a small number of surveyed enforcers.[3] One respondent put the matter precisely: "Violations would have to be egregious, repeated, and intentional with total disregard for the environment and public health."[4]

A small farmer expressed outrage at the use of the criminal law in his case:

> I was just completely taken [aback] with the type of handling. They [government enforcers] treated me like a criminal. They told us to go to [City X] and get printed and mugged. It was damned infamy for this area because it was me . . . and I'd say a half dozen other industry involved people. I'd say 10% of this area were in federal court getting printed and mugged like a bunch of common thieves.[5]

Many businesses consider criminal sanctions inconsistent with the regulatory problems which they confront. Criminal sanctions result in "overkill."[6] Business people use the term in two senses. Because criminal fines and imprisonment are excessive punishments, they should not be used. And, if used, these sanctions lead to attributions to the company of behaviors that are not involved in violations. Criminal remedies are thus unfair, since the public will make associations which devastate the firm's competitive profile.

Industry describes counterproductive outcomes of the criminal sanction. Industry will "overcompensate" or overcomply, negatively affecting the firm's attempts to innovate; industry will be chilled by potential criminal punishment from sharing ideas with others. "God knows what we've overlooked internationally because of fears of [criminally suspect] contacts across the industry," said one automotive executive. Reportedly chilled also are legitimate attempts to challenge regulations: "You can't ever test a regulation [under a criminal sanction threat]. You can't go to the judge and say 'EPA was exceeding its bounds'. . . . It's hard to imagine a case where the criminal sanction works well."[7]

In evaluating the effects of sanctions on compliance, a long-range perspective is necessary. Government can bring a criminal charge against a company whose operatives, if convicted, serve a sentence or pay a fine and then close down the company and begin business under a new name. This mobility is a

---

[3] Survey enforcers discussed in Chap. 2, p. 36.
[4] See Chap. 2, p. 36.
[5] Interview F. (To maintain anonymity respondents are referred to here by letter only.)
[6] Interview A.
[7] Interview X.

challenge to the deterrent effect of all sanctions in many areas of business crime; but violators of environmental law may be more apt to establish a new identity after imposition of a criminal sanction than those convicted of more commonly understood crimes. This resilience exists because many highly polluting businesses have low profiles within a community. Citizens may hardly know that the business—indeed the service itself—exists. What is more, executives of such businesses themselves may be "invisible" within the area.

Criminal sanctions may also interfere with cooperative government–business relationships. Business people overwhelmingly see themselves as law-abiding. Environmental law should promote information gathering on polluting activities and encourage compromises on societal goals. Criminalizing an act only gets in the way of business's social "contract" to behave in a reasonsable manner. Threatening to use criminal law obstructs the system of values that has evolved between the two sectors.

Scholz (1984), reasoning through game theory, advocates cooperative strategies because confrontation raises the costs of enforcement and compliance beyond a reasonable level, benefitting neither government nor industry. "At some level of stringency, compliance costs will become so high that firms preferring voluntary compliance at the previous level of stringency will now prefer taking their chances as evaders, imposing the higher costs of confrontation on firm and agency alike" (p. 281). The argument that use of criminal strategies promotes business evasion of law is partially valid (Scholz, 1984), yet can become a rationale to emasculate legal control of white collar crime.

Business also fears that criminal laws are misused once media descriptions of environmental damage touch public emotions. Present-day knowledge (e.g., about health effects) is applied to deeds performed years ago. A notorious recent manifestation of the problem is the *Johns-Manville* case; workers are bringing numerous suits against the company for health problems allegedly associated with exposure to asbestos. Practices considered standard and acceptable and indeed required by government during World War II are linked to asbestos-caused cancer and other diseases. A new group of environmental lawsuits also challenges common laboratory and construction uses of formaldehyde.[8] Businessmen feel that they are being held to an ever more stringent standard of care; criminal sanctions represent a new disincentive to entering the chemical manufacturing and use businesses.

Many business people associate criminal sanctions with deliberate life- or property-threatening activity and consider application to their actions as grossly

[8] Marc G. Kurzman and Judd Golden, "Formaldehyde Litigation: A Beginning," *Trial* 19 (January, 1983), at 82–85; "Insurance Law and Asbestosis—When Is Coverage of a Progressive Disease Triggered?" *Washington Law Review* 58 (1982), at 63; Anderson, Warshauer, and Coffin, "Asbestos Health Hazards Compensation Act: A Legislative Solution to a Litigation Crisis," *Journal of Legislation* 10 (1983), at 25.

unreasonable.[9] Only very marginal industries or marginal units within industries would behave so antisocially as to merit criminal punishment. And if subordinates do engage in illegal behavior to protect the company and/or their own jobs, management should not be criminally punished. Rather, the firm should have the flexibility to address the problem by firing these employees. Business executives are very concerned about criminal liability for employee behavior which they did not authorize.

Not surprising is the business preference for self-policing. But, in candor, some executives are realistic about public reactions to this notion:

> [Self-policing] . . . I think philosophically is a better system . . . I think [a] society which operates through a combination of self-policing with appropriate performance standards and periodic audits . . . is the most cost-effective system. . . . Fundamentally it depends on the honesty of the population. . . . (I realize though the) general belief that industrial sources are likely to be, if not generally dishonest, resisting all kinds of regulations.[10]

The suspicion is that businessmen consider those effective approaches which limit their ability to manipulate enforcement to be unacceptable. Industry opposition to the noncompliance penalty (NCP) suggests this conclusion. NCP is a provision of the Clean Air Act added in 1977 (section 120) to augment civil and criminal sanctions available under the act:

> Congress added section 120 to the Act because it anticipated that even the augmented civil and criminal penalty scheme would not create sufficient incentives for sources to comply with air quality standards. . . . Congress also hoped that the section 120 penalties would increase administrative flexibility in enforcement of the Act, by serving as a middle ground between stiff criminal sanctions or shutting down of noncomplying facilities. . . . Equally important, by removing the economic benefits of noncompliance, Congress hoped to place polluters on the same economic footing as those who had limited their emissions through increased anti-pollution expenditures.[11]

Penalties reflect the value to a source of noncompliance. The amount charged includes capital and operating costs and maintenance expenses which the firm avoids by failure to comply. To determine penalties the EPA developed a "mathematical penalty calculation model" which addresses benefits derived by violators and costs which a firm has incurred. Thus, under the NCP, government collects the amount of money that a firm has "earned" because of its failure to comply. Economic theorists have hailed the strategy as administratively simple, efficient, and fair, but the executive interviews describe the strategy in much different terms:

---

[9] The appropriateness of civil law because of its aim of correction, rather than punishment, was also noted by a few respondents to the enforcer survey (as discussed in Chap. 2, p. 36).

[10] Interview X.

[11] *Duquesne Light Co. v. EPA,* 698 F. 2d 456, at 463. For more on NCP, see Chapter 3.

> I think it should be repealed. . . . It's one of the stupidest laws I've ever seen. It's like
> handing somebody the death penalty for possessing marijuana. . . . It even denies
> authority to the courts. . . . it's one of those . . . principles that are caused by political
> concern of the moment.[12]

Individual companies and trade associations including Duquesne Light Company, Pennsylvania Power Company, the American Iron and Steel Institute, and the Chemical Manufacturers Association brought suit challenging the noncompliance penalty policy. Industry argued that the rules implementing the strategy are inflexible and do not allow the business sector to present its view of the noncompliance problem.[13] These assertions were rejected for the most part in the federal courts. Specifically the Court of Appeals for the District of Columbia Circuit found that judicial review of an NCP determination is available and exemptions exist which build in "flexibility" for sources which demonstrate that noncompliance is beyond their control or which meet other tests. The court concluded that it "is more than five years since Congress enacted this section (NCP); it is certainly time to put into operation the penalty assessment system Congress mandated."[14]

The regulated community vigorously opposes attempts to legislate greater use of the criminal sanction. The degree of opposition raises concern over efficient use of enforcement resources: employing criminal sanctions may raise a flag for industry and engender business responses to enforcement efforts based on principle rather than on considerations of cost–benefit analyses. What may be admitted and resolved quietly if classified as civil may become a *cause célèbre* if criminal, "like the acts of a mugger or rapist."[15]

Some enforcement personnel take a much different view of use of the criminal sanction. The Los Angeles district attorney described a case of deliberate dumping of toxic materials into the city sewers:

> There is only one way to deal with someone in a life threatening activity and this is
> jail. . . . If it is only a fine, it [the fine] would be considered part of the cost of doing
> business.[16]

Indeed, some commentators conclude that vigorously attacking the most flagrant violators, or what are called "the bad apples" (Kagan & Scholz, 1981), is essential to a meaningful deterrence posture:

> The worst case deterrence threat is what really determines the level of "voluntary
> compliance" that firms and enforcement agency tacitly agree on. Firms are not
> concerned with just the initial probabilities of detection and punishment, but with the

---

[12] Interview B.

[13] *Inside EPA*, Feb. 11, 1983, at 14.

[14] *Duquesne Light Co. v. EPA*, 698 F. 2d 456, at 486.

[15] Interview E.

[16] Ira Reiner, *Los Angeles Times*, May 26, 1982.

long-term probabilities that increase dramatically as the agency focuses its attention on the major violators.[17]

This assessment argues for at least some use of draconian responses to noncompliance in appropriate situations. These include criminal sanctions.

Cost–Benefit Calculations of the Regulatee

The small defendant may preceive the criminal sanction, at least at first, as something it must fight—regardless of costs—because of the destructive negative publicity associated with an indictment. Cost–benefit calculations can also take a back seat when an influential executive treats an enforcement act as a personal insult. Idiosyncrasies of CEOs, including views of their own expertise, can lead to reactions to prosecution that purely on the criterion of efficiency the firm would do best to settle. For example, the CEO may be particularly knowledgeable about the health effects of a chemical emission or of the Congressional intent regarding balancing health effects with economic factors. He or she may conclude that the interpretation given to a law by a citizen litigant, an environmental public interest group, or a government agency fails to reflect this knowledge. What in another area of law enforcement would be ignored as too trivial to pursue may consume an inordinate amount of time and resources—to fight a matter on principle.

Just as individual idiosyncrasies of top management can lead to a company's "digging in" and fighting implementation of environmental law, the preferences of top managers are an important factor in responsive behavior (Roberts & Bluhm, 1981). Executives may exert influence in a number of different ways. The CEO may have a personal commitment to environmental protection which translates into his subordinates' undertaking cost–benefit calculations in unique ways: What seems costly down the street may be calculated as the cost of doing business here. Or, regardless of personal preferences, the managers may establish controls which discipline company units to comply as standard operating procedure. (As we will see in Chapter 7, both the preferences and the controls may have less than perfect influence in highly differentiated firms.)

Despite individual differences, business does at times respond to environmental law in classical cost–benefit terms (Brown & Stover, 1977; Downing & Watson, 1974). Studies have shown that the costs incurred by the firm for noncompliance are trivial when compared with profits that can be made through violations. Marlow (1982, p. 170) concluded for OSHA:

If OSHA is to increase the injury control resources of firms above that generated in the

---

[17] Personal communication from John T. Scholz, June 19, 1984.

private market, it must increase the costs of noncompliance to the point of equality with the rates of return on alternative investments.

Former Los Angeles City Attorney Ira Reiner described the analysis done by a company charged with unlawful storage and disposal of PCB:

> There is a rhetorical question that must be asked: Why would an otherwise responsible corporation engage in conduct that is life threatening to the public? . . . In this case the answer is money.[18]

The case allegedly involved disposal by two unqualified machine brokers. The firm hired the men for much less than the bid of $43,915 it had received from General Electric Company to dispose of the material properly.

Thus, whether a sanction can contribute to an effective enforcement policy depends in part on government capacity to identify the point at which fines affect the competitive profile of a firm. No legal reasoning impedes setting fines in recognition of a firm's size and wealth, at levels that probably will affect the organization's cost–benefit calculations.[19] Some monetary penalties, however, will simply be added to the costs of doing business. Fisse (1980, p. 17) argues that a fine of even several million dollars to a company with annual sales of billions of dollars "would be tantamount to the payment of a tax or license." But there are limits to the private sector's ability to pass on costs; all firms have their "breaking points." Yet Stone (1981) concludes that, although "the law's money threats" may be superior to other sanctions in influencing business behavior, firms vary considerably in their sensitivity to threatened impacts on profit. Also, although fines at an "*in terrorem level*" may be legislatively available, it is not at all certain "that judges, prosecutors and juries will invoke them to the fullest" (p. 885).

Diver (1980, p. 266) has pointed out that in order to be effective "the regulator must estimate not the actual probabilities (of violation detection) but the probabilities as perceived by the regulated population." Other commentators (Likens & Kohfield, undated) note that the cost–benefit analysis is but a skeletal summary statement of the firm's forecasts, decisions, and responses. Diver (1980) and Fisse (1978) identify potential tort liability, business losses following adverse publicity, future litigation costs, and costs incurred in coming into compliance.

Stone (1978) suggests that corporate losses deriving from enforcement of health and safety laws are not of the same salience as other losses. Certain legal actions are seen as acts of God and are not added to the ledger of analyzed and budgeted costs. Further, stark cost–benefit thinking does not take into consideration the various ways in which business approaches risk taking (Stigler, 1970).

---

[18] *The Register* (Santa Ana, California), December, 1983, at A1, col. 2.
[19] *State ex rel. Brown v. Dayton Malleable, Inc.* 438 N.E. 2d 120 (Ohio).

There exists no common monetary measuring rod for preferences in the politics of regulation. As Wilson (1980, p. 359) concludes, profit maximization is "an incomplete statement of corporate goals."

> Enforcement, happily, is not the sole means of assuring compliance with regulatory directives. Business obeys regulations for a host of reasons—moral or intellectual commitment to underlying regulatory objectives, belief in the fairness of the procedures that produced the regulations, pressure from peers, competitors, customers, or employees, conformity with a law-abiding self image.

Business responses to the costs of coming into compliance vary. Had he known of the expenses of defending himself in an environmental lawsuit, a small businessman reported, he would have purchased the technology that could bring him into compliance: "It would have been foolish for me not to."[20] (Yet he decided against a plea bargain that would have cost one-half of the legal fees needed to pursue the case: "We are innocent and we didn't want to compromise."[21]) Another interviewee from a large corporation decided to enter a *nolo contendere* plea because it would result in a fine only one-seventeenth as large as the cost of pursuing a full criminal defense. "Certainly you sometimes will compromise some matter which you may think you're right on, simply because . . . protecting or preserving that position does not justify the costs."[22] Another company issued public statements that, because of its good compliance record, it would never plead guilty or "no contest" in an air pollution case. The suit involved alleged deterioration of neighboring residences and health effects linked to the firm's smokestack emissions. Nonetheless, the company later entered a *nolo* plea after the court admitted dozens of citizen complaints as evidence. The firm's attorney stated that, despite continued belief in innocence, "it's not worth the cost involved to prove it."[23]

Perceptions of cost also vary with other factors in the compliance framework. Activity by enforcers and support groups can influence the "base line" of cost calculations that the firm employs. As Roberts and Bluhm (1981) found:

> What appeared to be cheap or expensive to each organization depended heavily on their expectations, which changed over time. Decision rules and standard operating procedures adjusted to new circumstances; what had once appeared to be an enormous and inappropriate burden could begin to seem normal and routine. Particulate controls are a good example . . . Up to a point, they no longer seem unreasonable, unexpected, or impossible to bear, whereas twenty years ago they would clearly have been seen as all three. (p. 337)

We find sufficient examples of cost–benefit calculations in the toxic dumpings cases to support the conclusion that in some areas of environmental control

---

[20] Interview E.
[21] Interview E.
[22] Interview C.
[23] Interview H.

the "rational man" model may be useful for devising control policies. Some executives compute profits from illegal dumping activities, analyze the probability of being apprehended, and assess their ability to avoid penalties or to dilute the negative publicity of a possible conviction. After a raid on the Los Angeles-based Aaro Company, when Aaro personnel were apprehended in the act of discharging toxics into the Los Angeles city sewers, the city attorney described the company motivation: An Aaro executive did not wish to spend $7,000 a month to dispose properly of wastes into a permitted facility "when he could do it, in effect, for free by pouring it down the sewer."[24]

In thinking in cost–benefit terms, the target may take the long run into consideration. A company may seek a precedent-setting victory which would remove any threat of future prosecution. One auto executive described a low-probability defense to an EPA action this way: "Quite frankly, what we were looking for in litigating that particular case was kind of a no-harm, no-fault decision. It might have been a longshot. . . . We didn't get it."[25]

From the government perspective, business targets can affect the cost–benefit calculation by strategically manipulating legal procedure. A routine enforcement activity can be very costly. In *Oxford Mushroom*[26] EPA was unable to quash a subpoena of the administrator to testify personally in a distant city courtroom on his involvement in rule promulgation; the court concluded that the administrative burden and inconvenience were "slight." But slight inconveniences aggregated over several cases can lead government officials to step back from prosecution in all but very strong cases.

Innovative criminal and civil sanctions offered to address white-collar and corporate crime suggest that the nature of the incentive or penalty, rather than the nature of the law, may most strongly determine the deterrence and compliance-promoting power of regulatory enforcement. A series of civil suits[27] may be costly to a company's image. A well-publicized administrative order may be timely in a given economic environment. Conference and conciliation with the new regime of a firm also can be influential if professionally undertaken by top agency administrators. Perhaps less important than whether a punishment or incentive derives from a civil or a criminal cause of action are the costs which the firm associates with an enforcement approach, the behavior required of a businessman for noncompliance, and the attitude of the regulatee toward the overall regulatory policy. These are mediated by several characteristics of enforcement policy addressed below.

[24] Ira Reiner, *Los Angeles Times,* May 26, 1982.
[25] Interview D.
[26] *United States v. Oxford Royal Mushroom* (see Appendix for complete citation).
[27] One example involved a large chemical company after government had acted civilly against seven of its plants. A spokesman concluded: "There's a heightened awareness (of environmental matters) within the company." *The Wall Street Journal,* Sept. 29, 1981 at 48, col. 1.

Fairness, Legitimacy, and Rationality of Enforcement Policy

Although strong enforcement efforts in certain situations can produce compliance (Boydstun, 1975; Kelling *et al.*, 1974; Muir, 1967), perceived fairness of enforcement promotes business's willingness to comply and makes specific enforcement attempts more acceptable (Roberts & Bluhm, 1981; Rodgers, 1973; Skolnick, 1968; Stigler, 1970). Conversely, if sanctions are considered excessive, a businessman may assume that a law is not likely to be enforced or, if enforced, that it should be challenged (Argyris *et al.*, 1978; Levin, 1972; Packer, 1968; Sabatier & Mazmanian, 1978). Industry may take a posture of defiance, testing an enforcement policy in practice, if not legally. This reaction explains why some companies respond differently to civil and criminal sanctions. The deterrent objective of civil sanctions is more understandable to businessmen: it seems more reasonable than punishment linked to criminal sanctions.

In a sense, any enforcement scheme that business accepts might be labeled "fair"; all others are "unfair." One corporate executive admitted: "I think equity is basically a debating point because if I want to go and resist your proposed regulation, I will always use the equity argument, will always seek one, will always find one."[28]

But some objective indicators of unfairness exist. Industry cites as inequitable approaches to choosing of enforcement targets. The American automobile industry considers itself more vulnerable to suit than other pollution sources.[29] "Government attitude is the key. . . . We would like an even handed approach," one automobile company executive summarized.[30] The industry criticizes the EPA's implementation of the selective enforcement audit (SEA), used to monitor compliance with emission control requirements; American companies are reportedly subjected to SEA out of proportion to the numbers of vehicles they manufacture. EPA personnel are said to travel to Japan for audits only when the weather is bad in the United States.[31]

Other automobile company executives complain that the control of pollution should be analyzed in terms of costs to society; from the focus on the automobile industry gross inequities result:

> If there need to be control of these pollutants, surely it should be done at the least cost to the customers, to the public. You don't do that by squeezing automobiles right down to the very last drop of blood if you could have done it cheaper, such as by controlling electrical generating plants which also release the same substances. That is

---

[28] Interview B.

[29] Interviews A, B. See also *The Wall Street Journal,* June 6, 1982, at 1, 25: ". . . domestic automakers complain that it is inequitable that the foreign car makers get off so lightly," at 25.

[30] Interview X.

[31] Interview Y.

in fact what has happened. The automobile has been squeezed almost to the breaking point.[32]

Business complains when it sees enforcement linked to political ambitions, for example, a prosecutor trying to develop a national reputation by pursuing highly publicized environmental lawsuits. One automobile executive said:

> You garner more votes by attacking industry and appearing to uphold the little man against big industry. You have more votes that way than by taking a more reasonable approach.[33]

Stationary sources also argue the equity issue. The debate over plans to achieve federal pollution control standards in the air quality regions is a case in point. For example, the revised SCAQMD plan, announced in late 1982, admits that the Los Angeles basin will not be able to meet 1987 standards for ozone. Under one interpretation, the petroleum industry and the electric power generators would incur about 80% of the overall expense of clean up.[34] This allocation may seem reasonable because these industries do account for a large percentage of the air pollution problem. But while the plan was being developed elected officials in California were evading the federal requirement to impose automobile inspection and maintenance programs. The region has six million sources of pollution in the form of automobiles. Mobile sources account for 50% of reactive organic gases and much larger percentages of nitrogen oxide and carbon monoxide. In this context, business strongly questions the fairness of selecting individual industries for the most expensive controls.

Business may actively oppose an enforcement effort when it perceives sanctions as excessive. A Chrysler corporation spokesman called "outrageous"[35] a fine of $400,000 for what he judged to be a trivial violation. The company was also shocked to learn that another emission control violation could have resulted in penalties of $91 million.[36] Identical sanctions for grossly different violations also lead to criticism,[37] as do sanctions imposed after requirements that business considers infeasible or for which government does not provide technical assistance.[38]

The regulated industry bemoans lodging of both criminal and civil powers in the same agency:

> So that when you get yourself into a bargaining situation with an attorney over what

---

[32] Interview D.
[33] Interview Z.
[34] *Los Angeles Times,* Aug. 31, 1982, Part II, at 1, cols. 5–6.
[35] Interview (interviewee not indicated to preserve confidentiality).
[36] *United States v. Chrysler Corporation* No. 76–1800 (D.D.C., filed Sept. 27, 1976).
[37] Interview C.
[38] Interview E.

should be a simple matter, although he doesn't necessarily say it, you know he's got the brick in his hand that he can threaten you with and can use . . . bringing criminal sanctions if he doesn't like the way the civil case is going. There's only a very few areas in the Justice Department where they have combined civil and criminal [powers as in the Land and Natural Resources division. And that can lead to . . . substantial abuse.[39]

In some circumstances the failure to clearly communicate the enforcement strategy might pose a hardship for a firm. Industry has come to expect some notice of violation before action is taken, and this reliance has become part of the business–government relationship. In *Frezzo,* the plaintiffs argued that the EPA "must either give them some notice of alleged violations of the Federal Water Pollution Control Act, or institute a civil action before pursuing criminal remedies under the Act."[40]

The court decided otherwise but noted the existence of conflicting legislative history on the issue:

> Senator Muskie expressed the view in the Senate's consideration of the Conference Committee Report that an abatement order or civil action was mandatory under the Act. . . . A similar view was espoused by Representative Harsha in the House during debate on the House Bill.[41]

Nevertheless, the court relied on the final House Committee Report "which clearly indicated that written notice of the violation, administrative, civil, or criminal remedies under the Act were to be *alternative* remedies."[42]

What can be done to increase the sense of fairness of environmental law enforcement? Government–business contacts can foster a sense of equity. Some executives report that they are involved throughout the development of rules and of enforcement strategy, and they would consider it a corporate blunder if they learned of a regulation only after it was published in the *Federal Register.* Such persons appear to be more receptive to government activity. One vice-president described his corporation's intelligence work this way:

> So we would tend automatically to resist these [regulatory] ideas because of that inability [to know about them in advance] . . . We want to keep track of what's going on, what ideas are surfacing, how they're being surfaced, where they're being surfaced which means you have to know a lot of people . . . in the intelligence sense. Now that's the extreme leading edge.[43]

---

[39] Interview C; 42% of the surveyed enforcers, discussed in Chapter 2, p. 36, who are authorized to employ both civil and criminal sanctions reported that they were most satisfied with resolution of cases when a combination of criminal and civil sanctions was pursued as compared to 8% when criminal sanctions only were pursued and 25% when civil sanctions only were pursued.

[40] 602 F.2d 1123, 1126 (1979).

[41] 602 F.2d 1123, 1126 (1979).

[42] 602 F.2d 1123, 1126 (1979).

[43] Interview B.

Intelligence involves keeping informed of governmental activity at the federal, state, and local levels and monitoring corporate affiliate groups and union concerns and interests. Participation in the enforcement phase includes informal contacts during the period of development of agency strategies. Some regional EPA offices "would like to work with you to resolve what they perceive to be a problem . . . others tend to be much more like state troopers—you know, they've caught you in a violation and there's punishment."[44]

An expanding literature offers mediation of environmental disputes as one approach to fostering a sense of fairness in enforcement practices (Chapter 3). Although not all alleged violations are amenable to mediation (when, for example, noncompliance is egregious, deliberate, organized, and seriously health threatening, cooperative approaches may be inappropriate), mediation may effectively resolve many cases (Edelman & Walline, 1984; Patton, 1984). If properly orchestrated, mediation can avoid a firm's adversarial reaction to detection of (or speculation about) an environmental problem. Furthermore, mediation may conserve an agency's enforcement resources for more significant violations. Finally, interactive processes can promote communications between regulator and firm which generalize beyond the immediate problem.

Mediation can fit within an overall enforcement policy especially in situations in which possible disputes are recognized early. Parties to mediation can manage controversies in ways which avoid the formalization and rigidification of positions required in litigation (Susskind et al., 1983). In dispute resolution (as opposed to the more amorphous process of rule negotiation), parties with an interest are likely to come forward, and therefore in the enforcement phase the challenge to cooperative approaches of interest group identification is less disturbing. Thus the same cooperative processes which increase the firm's perception that government's environmental position is equitable may, in select cases, expedite dispute resolution.

Business also stresses rationality of the enforcement strategy. The dimension is subjective but regulatees consistently voice the theme. The firm criticizes some enforcement activity as irrational because the violation is de minimis. Government attempts to punish industry activity that has "zero public health consequences" are criticized as wasteful. Minor runoffs, temporary equipment failures, and small, inadvertent discharges of pollutants should be reported by the violator but not be subject to government action.

Government efforts are called irrational when enforcement activities are expensive, time-consuming and ultimately ineffective. One top executive described an air pollution action against his firm as "sheer lunacy." The case was resolved in the defendant's favor. The proceeding was counterproductive to the EPA's mission because it diverted scarce enforcement resources.

44 Interview C.

"Government conceives of industry as the fat cats."[45] Representatives of the automobile-manufacturing industry articulated this theme while questioning the rationality of certain enforcement activities. Regulators centered attention on the auto industry because historically car making had been quite profitable. The position has an ironic note. Industry respondents gleefully offered economic difficulties as proof that the effects of regulation had not been exaggerated by the corporate sector.[46]

Certainty and Imminence

Certainty and imminence of enforcement or reward foster organizational compliance (Chambliss, 1966; Ermann & Lundman, 1975; Gibbs, 1968; Horai & Tedeschi, 1969; Rodgers, 1973). These dimensions are not easily established; a variety of events can militate against a sanction affecting a target. Some inhibitors involve the agency's inability to secure sufficient evidence to make a case. Industry can be quite hostile to compliance monitoring. Opposition to OSHA inspection is legendary, but problems are not limited to the occupational sphere. Recently government employees have suggested that EPA enforcement agents carry weapons during inspections of alleged violations by organized crime of waste disposal laws.[47] In one year alone a dozen violent or close to violent incidents involving EPA inspection agents were reported, ranging from firing upon agents to releasing attack dogs on an air quality inspector.[48] On the site, the inspector may be inhibited or intimidated. An employee of an air quality management district reported:

> So I checked by again in about a month and they were smoking again [describing the stacks in a firm he had been inspecting]. So I went in, and unfortunately when I went in the guy had just received a letter for the last violation and a fine included, and he was already upset. He was fed up. So I was trying to be very careful with him. . . . It was a very sensitive issue. This time I was in . . . the small room where the guy was operating. I didn't want to stop him because the tires were spinning and it was very dangerous to get in there. So . . . before I could explain anything he just flew off the handle and threatened to kill me and he was going to knock me through the window, saying: "You can't tell me I'm not doing anything to stop the emission." He was doing it for his employees because he thought he was being belittled . . . and he started cussing. He said, "I don't have to sign that ticket" I said, "You don't have to," and I just laid the ticket down and said, "I'm not going to argue with you because you're obviously upset" and I left. But all the way out of the plant he's yelling and

---

[45] Interview D. See Frezzo's similar complaint in Chapter 1.

[46] Interviews C and B.

[47] In July, 1984, DOJ temporarily deputized EPA agents as U.S. marshals. *Wall Street Journal*, Jan. 7, 1985 at 14, col. 1.

[48] *The New York Times*, Oct. 26, 1983, cols. 5–6.

screaming at me and all the employees are looking, and he's calling me every name you can possibly imagine.[49]

Personnel movement within the company and diffusion of responsibility for the firm's actions can make threats of sanctions and promises of rewards meaningless. Prosecution can fall outside of the time frames according to which personnel operate, or enforcement efforts may not be directed to appropriate parties. Government often reduces the credibility of enforcement threats by extending compliance deadlines (Marcus, 1980) as was the case for electroplating operations which have consistently failed to meet standards under the Federal Water Pollution Control Act regarding discharges of heavy metals.[50] *The Wall Street Journal* reported that many companies were watching to see if the EPA would actually impose the very large threatened fines on General Motors for violating these rules. In particular, smaller firms were monitoring the EPA–GM negotiations "before they go ahead with large investments to clean up discharges from eloctroplating plants and other industrial sources."[51]

Numerous organizational deficiencies counter imminence and certainty of enforcement. These range from inadequate resources for surveying likely noncomplying sites to agency inexperience in law enforcement and in use of sophisticated testing equipment.[52] Some attempts to make enforcement more efficient may actually be counterproductive. Commenting on the failure of the government in pollution control cases to decentralize enforcement, one former assistant United States attorney described some of the costs of so-called coordination:

> The problem with criminal enforcement is that the [government] insists that the case bounce around for a couple of months [within the bureaucracy]. The right way to do it is to have the agent in the field work directly with the U.S. Attorney. This avoids the hassle of [a case] bouncing to [Washington] D.C.
>   Centralization makes good sense in a tax case, but not here. . . . It discourages good cases from being brought. You need to turn the prosecutor loose with an enthusiastic agent who wishes to make a case.[53]

Agency failure to monitor consent agreements also produces skepticism about the imminence of enforcement activity. Even cases of obvious fault and guilt take considerable time and enormous judicial resources to process. It may be years before the midnight dumper, "caught in the act," has exhausted all options within the legal system and finally has to serve a sentence or pay a fine. Litigation can postpone the final articulation of an order. Budget and staff limitations, absence of data, lack of interagency coordination, and agency reputations

[49] Interview H.
[50] *The Wall Street Journal,* June 29, 1984.
[51] *The Wall Street Journal,* June 29, 1984.
[52] *The Wall Street Journal,* Jan. 7, 1985 at 1 col. 6.
[53] Interview with Bruce Chasan Jan. 8, 1985.

of enforcement laxness all reduce the potency of regulatory threats (Clinard & Yeager, 1980).

The *Distler* case (Chapter 1) illustrates one message given by the criminal justice system. Recall that the defendant was a small businessman who was indicted on several counts of water pollution law violations for dumping of hazardous wastes. He received a record sentence in an environmental law case: two years in prison and a fine of $50,000. But it took over four and one-half years to end his legal appeals; by that time the defendant had developed a new reputation in his community. He had begun a successful business, one that required licensing by a state board that reviews the character of the applicant. Finally he did enter a minimum security prison and served part of the sentence.

Well-known aspects of criminal procedure partially explain the time period between violation and punishment. Whatever the merits of excuses, attributions of fault to the government, appeals, requests for rehearings, and other time-consuming legal maneuverings counter an image of effective government sanctions.

Surprise and "Compliant for a Day"

Publication of enforcement strategies that informs violators of the probabilities of being inspected, sued, or prosecuted also dilutes the perceived imminence of enforcement (Diver, 1980). Industrial targets can come into compliance for the period in which government inspects, or companies can otherwise alter behavior just enough to avoid prosecution and then return to noncompliant performance. The EPA decided in February, 1982, that "no enforcement action would be undertaken" when a violation "does not evidence intentional or repeated disregard for the law."[54] At one point in the early months of the administration of President Reagan, the EPA simply announced that, rather than using the adversarial approach that had characterized the past four years, the agency would "sit down and discuss the pollution problem with" industry. The aim would be to come up with a mutually acceptable schedule for achieving compliance,[55] not to identify violators and prosecute them. And in October, 1982, the agency publicized its even newer enforcement priorities for criminal cases: over half were to be aimed at hazardous waste violations under Superfund and the Resource Conservation and Recovery Act and at "willful contempt of . . . consent decrees."[56] The EPA did not guarantee companies not covered by these sets of environmental regulations that they would not be investigated; but the

[54] *Inside EPA* February 26, 1982, at p.8.
[55] *Inside EPA,* Nov. 13, 1981, at p.12.
[56] *Inside EPA,* Oct. 29, 1982, at p.12.

perceived probability of an enforcement action decreased with the agency announcements.

An earlier description of EPA enforcement priorities is another case in point. Overall enforcement efforts were to focus on: (1) conduct causing substantial harm or posing health hazards; (2) sustained and significant violations in circumstances in which compliance is clearly feasible; (3) falsification, concealment, and destruction of material records and information; and (4) willful contempt of civil consent orders. Polluting industries can reasonably conclude on the basis of these criteria that enforcement activity will center in the last two areas, where violations are egregious and government can identify targets through simple routines. A cynical interpretation is that such factor analyses are a way in which government can wink at industry while announcing its enforcement policy. A notorious example involved the regulation of lead in gasoline in 1982. Reportedly the EPA administrator, in response to a waiver request of federal rules which would allow greater lead content in gasoline, "did not grant the waiver in writing. But . . . she gave her word that she would not enforce the existing standards, and she encouraged . . . [the petitioner] . . . to ignore them."[57] According to a document written by one of the participants at a meeting with the administrator:

> We all thanked her and then left to meet . . . [the assistant administrator] . . . for a
> social visit. [One of the petitioners] . . . however, remained behind with . . . [the
> administrator] . . . momentarily. When he came out he told us that the Administrator
> explained to him that she couldn't actually tell us to go out and break the law, but she
> hoped that we had gotten the message.[58]

An environmental agency communicates that enforcement is neither imminent nor certain when it awards variances to noncompliers and extends deadlines to meet once strictly articulated standards. Or government may simply drop, as it did in 1981, dozens of enforcement actions, some because of statute of limitations problems.[59] These decisions do not go unnoticed. Subsequent to an extension of a deadline when manufacturers were to substitute water for oil in producing paint, the *Los Angeles Times* reported under the headline, "Smog Board Again Extends Deadline on Latex Enamels":

> Paint manufacturers have been given another two years to learn how to make good
> quality water-based interior enamels as a way to help reduce smog in the South Coast
> Air Basin.
>   Friday's action marked the seventh time since it first adopted air pollution con-
> trols on the contents of paint cans that the board of the South Coast Air Quality

---

57 Eliot Marshall, "The Politics of Lead," *Science* 216 (April 1982), 496.
58 *Ibid.*
59 "EPA, Citing State Deferrals, Asks Justice to Drop 49 Enforcement Cases," *Inside EPA,* Nov.
13, 1981, at p. 14.

Management District changed the rule. Had the rule not been amended, oil-based enamels would have been barred from store shelves and manufacturing plants next month. . . .

The Board left the way open to extend the deadline on enamels again in two years, if the industry shows that it still cannot produce suitable paints.[60]

Regulators often describe extensions as part of an innovative and flexible enforcement policy laid out in prescriptive terms, to make it appear as if no major change in enforcement approach is planned. Government will allow extensions only if a series of conditions are met, but the conditions may be no more than a rearticulation of industry's reasons in insisting it cannot comply. Consider the criteria imposed by the EPA for new compliance deadlines with air regulations by stationary sources:

Both the Justice Department and EPA's Office of Legal and Enforcement Counsel have concluded that a district court has equity power to fashion relief that allows a source to continue in operation beyond 1982 while taking steps to come into compliance . . . if . . . at a minimum all of the following threshold criteria are met: 1. the source must be unable to comply by December 31, 1982, other than by shutdown; 2. the source must demonstrate that there is a public interest in continued operation of the source which outweighs the environmental costs of an additional period of noncompliance; and 3. if there is any doubt about the source's financial condition, the source must demonstrate that it will have sufficient funds to be able to comply expeditiously.[61]

Although nonjudicious use of variances and extensions clearly counters industry's motivation to comply, it is not always clear under what conditions government should grant a petition for more time. In administrative hearings, often run like litigation, the complex trade-offs involved in implementation of environmental law are developed in great detail. Petitioner may argue that an extension now will buy greater compliance in the future. The company seeking the variance may assert that it is waiting for a new technology to enter the market and that a vendor promises new controls "within a short period of time." The petitioner will outline the effects of imposing controls immediately; these may include great monetary costs and layoffs that are linked to stricter controls.

---

[60] *Los Angeles Times,* Aug. 8, 1983, Part II, at 3, col. 2.

[61] Quoted in *Inside E.P.A.,* Sept. 24, 1982, at 3–4. The EPA's reasonable efforts program is of interest here. The program, not authorized by legislation, addresses EPA treatment of nonattainment areas where compliance with the 1987 Clean Air Act standards is considered impossible. Under the program, rather then bringing enforcement actions, the agency will review the control measures proposed by nonattainment air quality districts. The agency will share information about innovative control strategies used in other districts, and it will assist the nonattainment district to adopt rules which will move toward attainment in the long term. EPA will then undertake cooperative audits to determine if control measures are being implemented. Among the program aims is the avoidance of sanctions where reasonable efforts are being made to achieve air quality results. The program is controversial. See Chap. 6, pp. 117–118.

Agencies committed to maintaining original compliance deadlines will do so in the face of immense pressure for delays. These delays, considered individually, promise greater benefits with little cost.

Changes in enforcement policy militate against compliance. Industry dislikes uncertainties that derive from alterations in case settlement policy; yet inconsistency across administrations, coupled with knowledge of the time required for careful investigation and filing of charges, allows potential violators to estimate the probability and timing of a sanction's imposition.

Enforcement personnel changes with political administration, and these shifts weaken overall enforcement effectiveness. In 1980, the federal government was increasing use of criminal sanctions, but three months later the criminal strategy was set aside; the EPA now favored decentralization of enforcement and placed faith in approaches that rely on industry profit motivations. Substitution of conservative, business-oriented heads of enforcement for liberal, environmental activists within the major federal agencies leads to the conclusion that the federal role in environmental protection will decline. Subsequently, subordinate activist lawyers leave government. They are replaced with people who are more sympathetic to explanations of noncompliance or who, because of inexperience, do not fare well against industry lawyers. In this regulatory environment, the polluting business may contest enforcement activities that previously would be considered certain losses.

Despite textbook presentations of effective compliance systems, the unexceptional performances of regulators, government and business lawyers, and spokespersons also affect information flow. Mediocre performance is not a surprise in light of the demands on both regulators and the regulated. Both industry and government face turbulent regulatory environments; there is much to be learned and much that is not known in the field of environmental protection. Proceedings aimed at improving controls are often long and laborious, not because of malicious motives of those involved, but because of their inability to comprehend or fully to process relevant information.

The absence of machinelike processing of information is not limited to the administration of policy. Policy making also suffers from institutional inexperience. What appear as good ideas for getting compliance when considered in the silence of a professor's office or in an administrator's diary are often difficult to communicate persuasively to groups which must be won over. Use of incentives techniques is one area plagued by complexity. A SCAQMD Board member candidly lectured its advisory board:

> Frankly, despite the number of times you have presented the idea of full cost emissions charges, I doubt if more then three people in the room (of a Board numbering a dozen) knew what the hell you were talking about.[62]

---

[62] South Coast Air Quality Management District Advisory Council meeting, January 23, 1985.

Enter the Courts

Judicial proceedings can also counter general compliance as they contribute to an impression that enforcement is uncertain and slow (Marcus, 1980). Part of the delay in environmental law has to do with normal judicial proceedings and may not be greater than in other areas. For example, the average period from the time of filing a complaint to resolution of a case under the California Environmental Quality Act[63] is less than two years even when appeals are made; and cases are completed under the Michigan Environmental Protection Act[64] in well under a year.[65] But environmental law cases are not often expedited and the opportunity for getting continuances is great.

Use of innovative, ingenious, and time-consuming defenses is widespread. Consider again *Oxford Mushroom*.[66] The defendant argued that

> the Government circumvented the letter and spirit of the Act by failing to afford the Administrator the opportunity to become involved in a meaningful way in this case . . .

and that

> only through questioning Schramm and Costle may the extent of their apparent and actual control over the enforcement of the Federal Water Pollution Control Act and the initiation of criminal prosecution be learned.[67]

The United States had moved for a motion to quash subpoenas of Jack J. Schramm, a regional EPA official, and of the EPA administrator, Douglas Costle. The government described the busy schedules of these men, but the court, not moved by the agency's response, denied its motion. Requiring attendance by high-level agency personnel at individual judicial proceedings could effectively halt enforcement activity. It is unlikely that *Oxford Mushroom* actually decided that the testimony of the EPA administrator was essential to a fair trial. Rather, the litigation strategy may have been to make prosecution personally costly to agency administrators with the aim of encouraging a favorable settlement. (To be fair, another interpretation is conceivable. The defendant may have concluded that he was subject to criminal action for practices that, at the very worst, were slight aberrations from standard farming practices in his locale. A jail sentence was a possibility and a fine almost a certainty. Adverse publicity had already

[63] Cal Pub. Res. Code §§ 210001 *et seq.*

[64] Mich. Comp. Law Ann §§ 691.1201–.1207 (Supp. 1973).

[65] Precise statistics are not available. These estimates come from Sax and DiMento (1974); telephone conversation with attorney in enforcement division of Environmental Analysis group, State of Michigan Department of Natural Resources, October 27, 1983; and with W. Dickson, Legal Clerk, California Resources Agency, October 27, 1983.

[66] *United States v. Oxford Royal Mushroom* (see Appendix for complete citation).

[67] *Oxford Royal Mushroom,* Defendants Response to Motion to Quash Subpoena, at 3.

occurred. To call the "boss of the agency" who was suing the boss of the defendant company may appear quite logical to a small businessman. Such logic, in fact, explains a good share of business conclusions about the effects of regulatory law.)

A more common example of manipulation of judicial proceedings sees business producing for the regulatory agency reams of irrelevant information, ostensibly in support of the defendant's assertions. The strategy is legal but adds to the time necessary to reach a judgment. Requesting a jury can also postpone a trial, although in environmental law cases the use of a jury may not favor the defendant.

Judicial attention to due process and property rights is an important part of the American legal system that compliance goals should not overwhelm. Nonetheless, application of traditional understandings of these rights is often strained to protect white-collar defendants.

Studies confirm that the courts view white collar-crimes (of which some environmental violations are examples) as a special category. Mann, Wheeler and Sarat (1980), using interviews of federal court judges, conclude that in sentencing white-collar criminals judges are primarily interested in general deterrence. Judges consider the process of indictment to be punishment enough for the white-collar criminal, and they wish to limit the harm done to innocent parties, including the relatives of the convicted. Rather than addressing allegations on the basis of established facts and law, the judiciary often adds an element of compassion and understanding in environmental compliance suits—a factor which would be unusual if not bizarre in other criminal cases.

In *A–Z Decasing*,[68] a suit charging a battery company with polluting soils with toxic chemicals by improper production and disposal methods, the judge stated that he was "perfectly satisfied" that A–Z was "trying to comply" with the California Health and Safety Code. At a hearing on a permanent injunction, the defense succeeded in having the court delay for several months a decision on whether to compel a cleanup. The court hoped that the defendant could show that it had corrected the remaining violations. Similarly, in *Capri*[69] the judge, declaring that "the time for any danger to the community has long since passed,"[70] excused the defendant from probation, and dropped a contempt of court citation. The court considered whether a general outcome, the clean up of hazardous acids and contaminated soils, had been achieved rather then whether deadlines established for that outcome had been met. Originally, the company's owner had been sentenced to thirty days in jail, and a $3,000 fine had been imposed. The court suspended the sentence to allow the owner of Capri to clean up the site. The

[68] *A–Z Decasing* (see Appendix for complete citation).
[69] *Capri* (see Appendix for complete citation).
[70] *Los Angeles Times*, Aug. 21, 1982, Part II, at 7, col. 1.

lawyer for the defendant declared the judge's action a "complete vindication" of his client's behavior. Through such treatment, government passes a message to industry that, even in the 1980s, environmental violations are viewed differently from other crimes: the threat of sanctions, although forcefully made by a small group of environmentally oriented government officials, will not always result in true criminal punishment.

In certain situations the judiciary will also establish the burden of proof in ways that make it difficult for government lawyers to prevail. Moreover, judges are typically not familiar with evidentiary matters in environmental cases. Especially in the lower courts judges may lack the background to assess arguments on causation and proof in cases, for example, in which health effects of pollutants are at issue; when experts differ strongly about the relevance or meaning of a scientific study or about the cumulative significance of many studies; or when the intricacies of risk assessment are being laid out. Special attention is required to educate the judiciary on issues relating to probabilities of one or another harmful outcome, on uncertainties inherent in predictions of harm and about the wide range of understandings within the scientific community as to what is a health effect of concern. Effects that may manifest themselves in twenty to thirty years and may lead to other types of health problems are different from the more clearly binary events reviewed by the judiciary in other areas of law.

Delays in decision making caused by the complexity of evidence are not unique to the courts. But the decision rules employed by the judiciary and the procedural protections it provides postpone the impact on business of a government enforcement action.

Continuity and Consistency

To promote compliance the enforcement or incentive policy must be perceived as continuous (Schwartz & Orleans, 1967; Skolnick, 1968), that is, not subject to change with shifts in the economy or in administrative personnel. Business may judge continuity by reference to enforcement of a particular mandate, or it may evaluate the total program of enforcement within an agency or within government generally (Ericson & Gibbs, 1975). Business knows whether the legislature has ceased its oversight function, leaving follow-up to an indifferent, understaffed agency or perhaps even to one that is hostile to the law. Industry monitors whether the judiciary has interfered with an agency's approach. Whether a priority item under one administration will remain so under another is important intelligence for a regulated firm. The private sector will comply with many regulations only if government cares whether it complies.

A former EPA employee expressed the need for consistency:

> It is unreasonable to expect the "assembly line" [investigating and analyzing cases, negotiating a settlement or bringing suit] to work if it is regularly dismantled or

reorganized or if top management does not send consistent signals about its commitment to making the system operate.[71]

The ex-enforcement agent was complaining about interference with his attempts
to negotiate a settlement in the *Inmont* case,[72] which involved a hazardous waste
cleanup. After he found the company "suddenly" unwilling to respond to his
settlement offer, he learned that another EPA administrator had approached
Inmont with a different summary of the EPA's position: a $700,000 settlement
package and not the $850,000 figure he was using.[73]

Shifts in enforcement policy in the early years of President Reagan's administration were numerous. Enforcement procedures were drafted and redrafted
with changing emphases on cooperation, civil penalties, and selective criminal
enforcement. The government moved enforcement authority back and forth from
Washington to the regional office.[74] In assuming the role of enforcement counsel
at the EPA, a new unit director announced that his guidelines would "supersede
the policies and procedures issued by the (former) enforcement counsel . . .
which are revoked in their entirety."[75] The "old policies" had been announced
just six weeks earlier. EPA observers predicted significant additional changes in
policy once William Ruckleshaus assumed the office of administrator and again
when he resigned.

Industry finds ludicrous and unconvincing regular statements of a new
stringency or a new attitude toward enforcement. Just before President Carter
was defeated in his bid for reelection, the Assistant Attorney General for Land
and Natural Resources promised, "Now, however, I believe we stand on the
threshold of a significant change in the nature of environmental enforcement
litigation."[76] When businesspeople hear these grandiose statements about energetic approaches to promoting compliance they may be somewhat concerned, but
they also know that effective enforcement policy requires a commitment of
resources over time and more than the impassioned rhetoric of an individual in a
lame-duck political administration. To be sure, the Carter administration had put
much of industry on warning that enforcement of the environmental laws would

---

[71] *Los Angeles Times,* Apr. 3, 1982.

[72] *Inmont* (see Appendix for complete citation).

[73] *Inside EPA,* Apr. 9, 1982, at 8.

[74] See "Sullivan: Civil Cases, Penalties Cut, State, Criminal Actions Stressed," *Inside EPA,* Nov.
13, 1981, at 13, 14; and "Perry Shifts Policy to Give EPA Regions Power to Issue Superfund
Orders," *Inside EPA* Oct. 22, 1982 at 1, 9.

[75] *Inside EPA,* Apr. 30, 1982, at 11, 12.

[76] Testimony by James Moorman before the Senate Subcommittee on Environmental Pollution, May
24, 1979, quoted in N. Tennille, Jr., Remarks to ABA Natural Resources Law Section "Criminal
Liability Under Federal Environmental and Energy Laws: The House Counsel's Perspective,"
Denver, Colorado, Mar. 24, 1982, at 2.

be of high priority, but affected businesses also recognize the important function they themselves play in determining the longevity of any political regime.

Solomon and Nowak (1980) describe an impact of inconsistent enforcement policy in another regulatory arena. The initial success of the Federal Trade Commission's use of consent orders to control corporate behavior was threatened by modifications which greatly altered industry willingness to cooperate.

> The prospect of industry-wide regulation has dampened the receptivity of the targeted companies to the potential benefits of compliance with the consent order provisions. Instead, the companies have retreated to an adversarial position, certain that no good can come out of administrative intervention. (p. 140)

To be effective, an enforcement approach, perhaps any approach, must be perceived as having some staying power.[77]

On the practical side, businesspeople conclude that an agency which demonstrates a longstanding commitment to achieving compliance will withstand industry attempts to erode the agency's influence. Nonetheless, the effects of continuity derive from dynamics other than the practical. Continuity in orientation to achieving the goals of a regulatory program leads to a kind of social contract that industry enters with government. It is a contract that becomes background for business decisions and planning, although the private sector may seek changes at the margin through institutional means. In this sense, regulatory programs become another cost of doing business—the equivalent of complying with a well-known, although thoroughly disliked tax law.

---

[77] Gilbert Geis has offered an alternative view of the effects on compliance of inconsistency in policy under certain conditions. Absence of consistency prevents a firm from behaving in a rational and self-protective manner. In a sense, a target is forced to be on guard constantly, making compliance much more risky. This phenomenon may be more relevant for companies prone to avoid regulations in stable regulatory environments. Personal correspondence, May 20, 1985. And Sax has argued:

> We must put aside the dominating idea that the legal system is to be designed essentially to institutionalize stability and security. Probably nothing is more urgently required in environmental management than institutions for controlled instability. . . . The old idea of a stable and predictable regulatory agency, patiently negotiating solutions that will then be fixed and unquestionable for years, or even decades, is hopelessly outdated. A mixture of legal techniques—designed to destabilize arrangements that have become too secure—is precisely what is needed for a milieu in which rapid change is the central feature. (J. L. Sax, A General Survey of the Problem, in *Science for Better Environment*, Ed., Science Council of Japan, 1976, pp. 753, 755–756. Quoted in Anderson, Mandelker, and Tarlock, Environmental Protection: Law and Policy (Boston: Little, Brown, 1984).

CHAPTER 6

THE BEHAVIOR OF COMPLIANCE

Communicating Law

> *It also is totally unrealistic to assume that more than a fraction of the persons and entities affected by a regulation—especially small contractors scattered across the country—would have knowledge of its promulgation or familiarity with or access to the Federal Register.*
> —*Justice Powell in* Adamo Wrecking Company v. United States, *434 U.S. 275, 290*

> *Two things you never want to see being made are sausage and legislation.[1]*

> *The refutation [of a probability] is not always genuine; it may be spurious: for it consists in showing not that your opponent's premise is not probable, but only in showing that it is not inevitably true. . . .*
> —*Aristotle,* De Rhetorica, *Book II, Chapter 25, 1402b (quoted in Engelberg, 1981)*

In a liberal democracy, enforcement by itself, no matter how fully supported and professional, cannot and should not insure compliance with law. Poorly conceptualized, badly drafted, and incompletely articulated regulations counter positive response to environmental goals.

Aspects of the enactment and promulgation of law and its subsequent communication to business influence compliance. The clarity and perceived rationality of the legal mandate and the consistency of its articulation are particularly important dimensions.

Clarity and Specificity of the Law

Ideally, the law will be understandable to its business targets and specific enough to inform industry what is to be done and what is to be avoided. The

---

[1] Interview A.

overall intent of the regulation should not be lost among its provisions (Dolbeare & Hammond, 1971; Krislov, 1972; Lowi, 1969; Rodgers, 1973). Clarity of purpose and clarity regarding the means of achieving compliance are both important. Congress, state and local legislatures, and administrative agencies often fail to meet these goals.

Fundamentally, the law must be so communicated that business knows it exists and what it means (Wasby, 1976)—a situation not always the case in environmental law. To be sure, ignorance of the law is an excuse that defendants will offer widely, and a regulatory policy cannot be continuously responsive to expressions of lack of knowledge of the rules. Nonetheless, despite elaborate command and control and permitting processes, gaps do exist in our system of law making. Definitions are sometimes extremely complicated, deadlines left unspecified, standards not noted. Or regulations aimed at implementing statutes may conflict with the language of the law. For example, in *Frezzo,* during the appeal process, the government had argued that the plain language of the Federal Water Pollution Control Act Amendments covered the alleged violating behavior. The court of appeals decided, however, that in a criminal case "petitioners were entitled to rely on the language of the regulations."[2] Communication deficiencies leave even the company with the best intentions unaware of at least some of the rules that affect it. An effective compliance policy acts to fill these gaps.

Business must understand statutes, ordinances, regulations, and judicial opinions in order to comply meaningfully. Diver (1983, p. 69) discusses the "transparency" of law and states that rulemakers should "use words with well-defined and universally accepted meanings within the relevant community." Too often, this is not the norm in environmental law. Nor are regulations generally accessible, that is, "applicable to concrete situations without excessive difficulty or effort" (Diver, 1983, p. 67). An Allied Chemical Company executive compared review of regulations for the chemical industry to exegesis: "It's like learning the Koran."[3] Indeed, an army of lawyers whose main function is to interpret environmental laws now appears essential in attempts to implement environmental policy. Employing this professional contingent may substitute for overall comprehension of law in the business community, but lawyers are trained to react to law in ways that do not always favor compliance.

Coffee (1981, p. 453) has commented:

> Consent orders . . . frequently consist of vague language which parties can legitimately read differently. . . . The Allied Chemical consent degree . . . required only that the corporation investigate environmental risks and take "appropriate" action. It is difficult to envision a court finding that such a vague requirement had been violated where the corporation made even the slightest effort to comply.

---

[2] 642 F.2d 59, 61 (1981).

[3] *The Wall Street Journal,* July 20, 1984, at 46.

Coffee's observation applies to environmental regulations in general. The regulated must achieve a delicate balance in communicating law; rules will fail if they are either vague or overly specific. The generality or ambiguity of the legislation that is the basis of regulation may itself be at fault. Legislators' compromises, interest group pressure, and advocacy of ambiguity by potential regulatees all frequently contribute to legislation's impotence.

Political scientists explain why ambiguity characterizes some law reform (Jones, 1977; Lindblom, 1959). If clear articulation of rules and standards were required at the legislative level, activation of numerous interest groups might act to thwart any reform or lead to dilution of regulatory action. By keeping parts of the control strategy unspecified and granting some discretion to administrators, legal change can bypass scrutiny by control targets.

But environmental law can communicate sufficiently without detailing all activities required or precluded. The law of public nuisance is an example: In *Narmco*,[4] a California court addressed the alleged vagueness of this cause of action, an important part of pollution control efforts.[5] Narmco's contention was that the state public nuisance statute[6] governing air pollution emissions is unconstitutional because "air contaminant," "detriment," "health," and "natural tendency to cause harm" are not narrowly enough defined to inform business of the standard to which it will be held. The *Narmco* case involved misdemeanor charges for chemical emissions from the company plant in Costa Mesa, California. These discharges allegedly resulted in offensive odors and a variety of health problems for neighbors of the company. The court held that the asserted vagueness of nuisance law was not a valid defense to the company's actions, of releasing smoke, dust, odors, fumes, and gases into the neighborhood air. The company pleaded *nolo contendere* in the face of vigorous prosecution and strong citizen complaints and was fined $500.

A risk of detail in regulation is that even a comprehensive statute may not specifically mention some actions which the legislature intends to control. Companies may thereby find loopholes. Articulation of all potential violations in all definitions of compliance is not possible. As Kagan (1984, pp. 3–4) put it:

> No maker of protective rules can fully envisage the diversity of technologies and the countless ways things can go wrong in a complex and dynamic modern economy, or the inexhaustible capacity of workers and managers to slip into previously unspecified modes of inattention, stupidity and heartlessness.

Specificity may also counter a desirable flexibility in implementation of law. Detail may prevent an administrator from wisely interpreting the overall message

---

[4] *Narmco* (see Appendix for complete citation).

[5] See Chapter 3 in general on the importance of nuisance to the evaluation of environmental law. See also Rodgers (1977), Chap. 2.

[6] California Air Pollution Public Nuisance Law, Health and Safety Code, Sec. 41700.

of legislation, leading to an outcome that the regulator opposes and that sets industry against responsive action. Some generality in a mandate may promote compliance, at least in the early stages of articulation of a policy (Coombs, 1980). Case-by-case application with some administrative discretion promotes the societal learning necessary to foster further understanding of a control challenge.

Because of the controversial nature of some environmental initiatives, the legislature, administrative agencies, and the courts often attempt to avoid specificity of mandates. The courts have rationales for avoiding articulation of standards, encompassed in doctrines of separation of powers, limited judicial review, primary jurisdiction, political questions, and others. Administrative agencies can rely on the nondelegation doctrine which asserts that all legislative power is lodged in the legislature and that lawmaking cannot be delegated away. In practice, the history of much environmental and social legislation is one of finger pointing and shifting of responsibility for vagueness, and lack of clarity of laws and rules.

An example of this unproductive relationship between those who enact laws and those who implement them evolved over the meaning of "how clean is clean," a phrase used to summarize an issue under Superfund in the cleanup of hazardous waste sites. Summarizing the performance of government officials, William Hedeman, the EPA director of the Superfund, said that standard setting for cleanup had been, like several questions under Superfund, "shunted from the legislative to the executive to the judiciary branch."[7] Citizen concern over hazardous wastes, described as "hysterical," had "prevented common sense from prevailing and caused policymakers to abandon responsibility"[8] so that the cleanup campaign lacked "direction that it sorely needs."[9] In the absence of Congressional direction, EPA officials differed dramatically on how to proceed. The debate between the agency's Solid Waste and Emergency Response Office and the Policy, Planning, and Evaluation Office was particularly heated.

Participation in Rule Making as an Avenue to Clarity

Properly channeled business participation in the development of law can increase the probability of realizing compliance. Participation can promote clarity of law and advance other attributes of acceptable regulation.

Chapter 3 presents several forms of participation by targets of environmental law and policy (Susskind & Weinstein, 1982, 1983). Rule negotiation (Bacow & Wheeler, 1984; Russell, 1985; Sullivan, 1985), mediation, analysis of

---

[7] *Inside EPA*, May 18, 1984, at 7.

[8] *Id.*

[9] *Id.*

environmental problems in workshops, informal communications at conferences, and courses on regulatory law can all assist in the process of communication of law.

These participatory forums have several characteristics which hold promise for improved rule making. Basically, interactive processes can identify the people and groups who care most about environmental regulations, clarify the aims of rules, determine trade-offs among objectives sought by interest groups, and set priorities on goals. These sessions may conclude that there are different acceptable ways of getting to the same end. And some forms of regulatory negotiation also allow more time for sharing of information which is the basis for rules—an advantage over working in legal forums where time pressures may prohibit understanding of background data and concepts.

Studies of participation in policy making stress several different lessons and caveats for promoting compliance. An extensive body of literature indicates that participation in rule making increases satisfaction with resulting laws (DiMento, 1976). Analyses of the dynamics that lead to decreased resistance to change vary, but some studies focus on the opportunity to specify the meaning of change. Others emphasize the commitment which evolves when the participant feels that change is not imposed from the outside.

Two opposite cases highlight how participation produces greater acceptability of environmental law. A negative example comes from that class of small business which has limited impact on the regulatory process. One interviewee complained about the inability of mushroom farmers to influence regulations, and he was decidedly unsatisfied with the regulations under which he was prosecuted. He characterized his affiliate group, the American Mushroom Institute, as ineffective in describing to lawmakers the problems that farmers face in meeting water pollution control standards.[10]

An opposite reaction is that of a large metal extraction company heavily involved in the development of regulations: "We would consider it a total failure if the first time we saw a regulation was in the *Federal Register*."[11] The company participated in rule development in several ways. Its trade association did formal lobbying in Washington, D.C., to educate legislators about the "special needs" of mining.[12] The company identified environmental and other regulatory issues before they were taken up by the legislature and made efforts to monitor public concerns over pollution in the mining industry. The ideal was to anticipate legislation and then act to mold or control it. The company also actively participated in the regulatory activities of the Clean Air Forum and the Business Roundtable. Aware of congressional sensitivity to employment impacts, the

[10] Interview E.
[11] Interview B.
[12] Interview B.

company repeatedly emphasized these in lobbying. Its executives also recognized that involvement must be continuing to be effective. Arthur D. Little and another consulting firm were hired to monitor the technical literature on health effects of air pollution. An outside law firm then coordinated this information in addressing proposed regulations. The firm organized its data according to categories of clean air law: welfare, health and visibility effects, and acid rain.

Ongoing involvement also can increase a company's receptivity to resulting rules. One automobile executive described the aim:

> Now the interpretation of the regulations is important. We try to tell the agency how to interpret, that there is flexibility.[13]

Sophisticated company representatives will act to create the context in which rule interpretation is undertaken. Influence can extend to the analysis of scientific information that precedes rule making. An example involves the relationship between oxides of nitrogen and ozone in the Los Angeles basin. General Motors has vigorously addressed the issue—one with significant regulatory implications, most notably for automobile emission systems. GM argues that lowering of nitrogen emissions may actually increase ozone levels. The debate within the scientific community on the relationship between levels of emissions and ozone pollution is complex. A large automaker's entry into the analyses can have two results: (1) pushing science, thereby expanding the knowledge base, and (2) increasing the chances that resulting rules will be acceptable to the automaker, regardless of the industry's impact on the creation of good science. (Later in this chapter we address the issue of the use of science as a "weapon" used by industry and others in the regulatory process.)

Participation will not always generate consensus on regulatory goals and means. But consensus need not be the aim; in some situations involvement in lawmaking can produce a greater acceptability of the regulatory process and consequently lead a firm to conclude that resulting regulations manifest characteristics of acceptable law: clarity, fairness, and rationality.

Participation in Rule Making as an Avenue for Regulatory Capture?

If not properly channeled, involvement of the firm can also lead to trivial compliance, as described in Chapter 2. If government does not control participation, resulting regulations may be overly responsive to industry interests. At the extreme, certain forms of industry involvement in rule making may result in the capture of the regulatory agency by the regulatee.

Capture is a phenomenon much discussed in the regulation literature (Bernstein, 1955; Kolko, 1965; Landis, 1960; Lowi, 1969; MacAvoy, 1965; Mitnick,

---

[13] Interview A.

1980; Noll, 1976; Posner, 1973; Sabatier, 1975; and Stigler, 1971). The general notion is that regulated firms, through a variety of means, can so influence a government agency that the agency acts to protect or promote the industry rather than to control it. Scholars differ on the nature of influence: some conclude that it is based on appointment to agencies of people who spend most of their professional lives in industry. The revolving door (the same people move back and forth from the private sector to the regulatory agency) and career succession (industry influence derives from concern by regulators over their future job prospects) are competing statements of the capture problem. Other scholars tie influence to budgetary incentives that exist to counter control of regulatees, to government deference to the experience found in the private sector, and to a bureaucratic tendency to seek information where it is offered, as opposed to generating new data and studies. The latter point converts to industry's bestowing on regulatory agencies considerable assistance in background work for regulations and for the drafting of laws.[14]

Little good empirical support of the capture theory exists, and its import varies with governmental and political form. Braithwaite reports that the phenomenon is less of a problem in Australia where elected councils exist for all major interest groups, including the consumer movement, the labor movement, and the trade unions. Thus the state guards against capture by "scrupulous counter participation by public interest groups."[15] The significance of capture also varies with the form of participation used. For example, in regulatory negotiation which emphasizes openness to those who select themselves, where parties are granted an equal footing and where the outcomes are *ad hoc* and time-constrained, the problem may be nonexistent. On the other hand, when collaboration is not publicly visible and government–business connections are longstanding and lack a previously determined end point, influence which undermines the rule of law may be more common.

Among the criticisms of capture is Quirk's (1981) study of federal regulatory agencies, which found little evidence of the phenomenon. Other studies have noted that the powerful American corporation is not fully able to dominate regulatory law. McCaffrey (1982) offers three reasons why corporations remain true targets of regulation, rather than molders of regulatory policy. First, unlikely proregulation coalitions evolve within the private sector because of divisions within and between regulated industries. Second, as industry lobbying ceases, the regulatory agency reacts accordingly:

> The responses of agencies to pressures do not reflect the balance of interested organizations because (a) the marginal values of technical lobbying decline significantly after a point and (b) regulatory organizations do not react in neutral ways to competing claims. (p. 404)

---

[14] I am indebted to Gilbert Geis for this latter point.
[15] Personal communication from John Braithwaite, July 2, 1984. See also Braithwaite (in press).

A third factor has to do with the forums for judicial challenge to regulations in which the powerful industries must operate; for example, federal courts differ in their treatment of regulatory initiatives, and the views and dispositions of some may well offset the resources of the better endowed litigators.

Despite this absence of an empirical basis, many social scientists (in particular, political scientists) and lawyers continue to warn of regulatory capture. The fear, which derives from a model of an adversarial relationship between business and government, is that business involvement will lead to control of agencies, to weak laws, and to government's acting overresponsively to special interests. Occasionally theorists will recognize that government can maintain a controlling position when cooperation is fostered, but predictions of government influence on industrialists are the exception. Even some of the mediation literature points to the immense resources of the private sector as a warning to advocates of consensual approaches to rule making and decision making on environmental issues (Amy, 1983). However, Susskind *et al.* (1983) and others describe ways of structuring negotiation and mediation that put government and public interest groups on a par with business. Public funding of consensus-seeking efforts is one equalizer.

Regulatory reforms which call for greater business involvement and participation must recognize the fear of capture—a fear made the more salient because political scientists and lawyers dominate the analysis and implementation of regulatory law. What is more, sufficient opportunities exist for industry to control the regulatory process to suggest that we must evaluate the benefits of clarity and acceptability of rules with knowledge of the costs of involvement.

Consistency in Articulation of the Law

Law can be clear and appropriately specific but fail to promote compliance because of inconsistency in its articulation. Especially under conditions of uncertainty, social influence, rather than universalistic statements or general criteria, affects decision making in complex organizations (Pfeffer *et al.*, 1976). In environmental policy, the political environment is typically highly uncertain; without consistent communication of law, the business target is not convinced that the same standards will apply over time. Administrative changes, personnel movement, and shifting societal priorities predict otherwise. In such situations, constancy of governmental policy fosters compliance.

Several studies of policy emphasize consistency (Krislov, 1972; Schwartz & Orleans, 1967; Skolnick 1968). An investigation of the compliance of lower courts with Supreme Court law in the area of libel (Gruhl, 1981) concluded that it was consistency, more than clarity of a policy statement, that had the greatest influence on lower courts. Clarity was

> perhaps . . . not as important in producing compliance as the variable of consistency. This suggests that the Court may be able to achieve greater compliance by handing down a consistent set of decisions rather than only one or two clear and unanimous decisions which provide the same rules. (p. 518)

Consistency in articulation of environmental policy is generally not the rule. Examples are numerous of changes in regulatory statements within administrations and, more dramatically, across administrations. For example, major changes discussed or implemented during the first years of the Reagan administration included alterations in the sewage treatment program, the waiving of several air standards, the easing of rules to allow firms to trade off pollutant allotments, delays in hydrocarbon and carbon monoxide standards for gasoline-powered heavy trucks, the overall revision of secondary treatment rules to relax effluent limitations, changes in the concept of the water bubble, termination of enforcement of noise regulations, new rules for liquids in land fills, changes in incinerator regulations, shifts in financial responsibility for hazardous dumps, alterations in consolidated permit reporting requirements, changes in lead exposure standards for battery manufacturing plants, alteration of strip mining rules and considerable discussion of the complete revision of the Clean Air Act—to choose among hundreds of items.

Concern over inconsistency comes from all sectors. Automobile executives have vociferously criticized the increasingly stringent emission control standards, arguing that it is difficult to achieve quality in control technology in the ever-changing regulatory environment.[16] Engineers in the stationary source division of some auto manufacturing companies considered federal regulations—as opposed to state implementation plans and programs—to be quite stable and expressed less concern about changeability of environmental rules than those responsible for mobile source emissions. Presumably the regulatory environment affecting stationary sources reflects a greater understanding of reasonable limits on pollution controls than is the case for mobile sources. This situation may change with pressure to develop a national policy on acid rain. However, attention to technology forcing for motor vehicles already reflects perceptions that translate to even more stringent standards: automobile manufacturers appear able to make changes related to style and aerodynamic features and thus should be equally creative in areas of health and safety. And environmentalists characterize the industry, despite periods of recession, as immensely profitable and capable of investing in air quality.

The EPA hazardous waste program has been another focus of criticism. According to the Environmental Defense Fund, the "on again, off again" requirements are said to be detrimental to industry as well as to the public. Lack of stability "certainly doesn't benefit the public . . . because of untimely delays,

---

[16] Interviews C and D.

massive confusion and no regulations"; industry "wants to know what the requirements are" for planning purposes.[17]

Some industry executives seek consistency and predictability in regulations even more than they promote more favorable regulations. Concern for stability elicits some surprising responses to deregulation initiatives. When former Secretary of the Interior James Watt announced pro-industry policy changes, *The Wall Street Journal* reported:

> The suits [by environmental groups challenging Mr. Watt's policies] already have had a chilling effect on industry executives looking for predictability in federal regulations.
>
> "You can't run a department if the courts are constantly peering over your shoulder," says a mining industry lobbyist who supports Mr. Watt. "It doesn't matter how much legal horsepower Interior has. The delays and uncertainty are terribly damaging."[18]

Business may seek regulatory consistency because it maintains competitive advantages over those who argue for easing of rules. In 1983, both General Motors and Ford sought changes in the fuel efficiency standards for their 1985 automobiles, but Chrysler Corporation desired to maintain the regulations since it was able to comply with them.[19] Deregulatory actions that may aid an industry as a whole may be detrimental to individual firms. Rule changes also raise questions of equity if some companies have made investments to comply while others have taken chances.

In the early 1980s, concern for consistency became clear in an ironic way when industry took a position favoring extension—rather than weakening amendment—of the Clean Air Act. The alternative was to support a Reagan administration bill that reflected greater sympathy to utilities and other sources of industrial pollution. The decision reflected fear of the unknown and the calculation that a more environmentally sympathetic Congress might be elected, should the old act not be continued, and impose an even more demanding law.[20]

Industry, most notably the Business Roundtable, has even opposed the self-auditing program created to respond to time-consuming inspections and command and control regulation.[21] Concern exists over evolution of the concept across administrations; it "might shift from a voluntary approach to a mandatory regulatory program."[22]

---

[17] *Inside EPA*, Nov. 12, 1982.

[18] *The Wall Street Journal*, Feb. 18, 1983, at 15, col. 2. Interview C noted: "If we're going to fix the process someway we need . . . to start maintaining some consistency with regard to the regulating approach so we may develop a cadre of people (in the agencies) who know the industry they're regulating and can live there long enough so that they understand it."

[19] *The Wall Street Journal*, Dec. 14, 1983, at A-1, col. 6.

[20] *Inside EPA*, Oct. 29, 1982, at 14.

[21] See Chap. 3.

[22] *Inside EPA*, Sept. 10, 1982, p. 3.

The industry argument for consistency in regulations is made disingenuously at times. Companies obviously oppose continuation of stringent standards and regulations deemed particularly detrimental to overall business performance. Selective enforcement is an example; it is a program based on emission testing of a sample of automobiles as they come off the assembly line (see Chapter 1). During the Ford administration, attempts to dismantle this approach to monitoring compliance were numerous. An EPA official reported that industry sought "loophole after loophole in the regulation and to exploit such loopholes. . . . The only gainer from an emasculation of the certification program will be those auto companies who want to get rid of as much effective control over the way they build their cars as they can."[23]

Inconsistency as Organizational Outcome

Knowledge of organizational behavior helps explain inconsistency in regulations. News coverage of environmental policy may focus on a singleminded government effort, which through rational means has identified a violator. Subsequently government will use its resources efficiently to promote compliance. However, government's environmental activity derives from a group of quasi-independent units which often have quite different organizational objectives. For a given case, differences may arise over whether there actually is a significant violation; whether the violator should be pursued; which strategies should be employed to prosecute or litigate and which sanctions work best.

Bureaucratic process produces results which may deviate from outcomes which legislators and regulators contemplate. Conybeare (1982) has noted:

> The organizational environment literature suggests that a complex regulatory environment will lead to a complex organizational structure in the regulatory bureau [fn. omitted]. Complex structures may inhibit the ability of the organization to follow a coherent regulatory goal, for reasons which have been collected in the literature under the term "bureaucratic process." H. L. Wilensky, and later Allison, emphasized the extent to which standard operating procedures for processing information and making decisions may inhibit the ability of the organization to respond to problems with appropriate policy. In the extreme, the policy output of the regulatory agency may be little more than a negotiated cluster of routine patterns of behavior bearing little resemblance to the original regulatory goals. (pp. 36–37)

Regulation of lead in petroleum is a particularly interesting case. Throughout the early 1980s the Reagan administration signaled that it would review the lead standard for small refineries. At issue was the amount of lead that can be included in nonleaded gas. Small refineries believed that a lenient standard would apply to them. But the final rule differed greatly from the original EPA

---

[23] *Environment Reporter* (1976), at 652, 653.

proposal. Pressure from larger oil companies may have influenced the change. Big companies oppose a differential standard because they are able to meet the more stringent rule with comparatively little difficulty. Small refineries had to alter their plans in response to the economically significant, if not devastating, change. The resulting rule was a disappointment to many within the administration and even within the EPA.[24]

Goal displacement occurs in environmental bureaucracies just as it does in many large organizations. Maintenance of the agency, rather than the realization of legislated objectives, becomes a main aim. Of course, commitment to environmental protection varies among agencies charged to be stewards of resources, but considerable activity is commonly dedicated to staff benefits, interagency posturing, and individual career building.

In some situations, compliance outcomes are accommodations reached, perhaps tacitly, between government and business. Vaughan (1982) writes:

> Since the information and wealth possessed by organizations can create obstacles to enforcement activities, agencies frequently fulfill their responsibilities through negotiation, internal proceedings, informal hearings, and mutually agreeable solutions. And business firms, similarly concerned with successful operation, soften the power of agencies by efforts to influence law-making and as a consequence, the nature of enforcement, and find equivalent gains to be had from negotiation. Compliance emerges as a product of the power-meditating efforts of both parties, as compliance demands fewer resources from both agencies and business firms than do adversarial activities to impose and thwart punitive action. In any given case, of course, a firm or agency may funnel all available resources into a full-fledged adversarial proceeding. (p. 1384)

The regulated bemoan lack of coordination among regulatory agencies and argue that it is impossible to respond to inconsistent regulations. Industry exaggerates inconsistencies for its own purposes but many observers note examples. A respondent to a 1980 Conference Board Survey (Lund, 1977, p. 47) concluded:

> Our major problems arise from the inability of the agencies to coordinate their activities adequately within their own groups and with external groups. I speak particularly of interaction between such groups as the Environmental Protection Agency and the Corps of Engineers, as experienced during our attempts to develop impact statements for our rolling mills project, and interaction between groups such as the Corps of Engineers and the U.S. Fish and Wildlife Service. We recently were involved in a permit application for the construction of a new storm waterholding pond for one plant. Because of differences in lack of communication between the Corps of Engineers, which had approved the permit, and the U.S. Fish and Wildlife Service, which had questions concerning some of the plans, the result of this protracted discussion was a delay of six months in receiving our permit.

---

[24] See also the description of internal differences within EPA over the draft benzene hazardous air pollution regulations reported in *Inside EPA*, May 18, 1984, at 14. The EPA's policy staff heralded the OMB position; the Air office seriously opposed the cost–benefit conclusions.

The EPA's differences with the Department of Justice over environmental lawsuits during the Reagan administration were widely publicized;[25] and in 1982 the Department of the Army was involved in another policy dispute with EPA. The army concluded that guidelines covering the dredge and fill permit program under Section 404 of the Federal Water Pollution Control Act Amendment should be strictly technical. Rules should not apply broad environmental policy of water quality enhancement. The two agencies differed on their interpretations of terms, including "waters of the United States," information required before permits are issued, and presumptions against discharging in wetlands.[26] EPA differences with the White House Office of Management and Budget (OMB) are common and in some cases industry can rely on OMB to postpone, if not ultimately weaken, standards.[27]

Coordination across bureaucracies responsible for regulating and enforcing is even more difficult during periods of political transition. Consider the situation at the EPA when Anne Gorsuch assumed her post as administrator. Although new employees in the enforcement division of the EPA may have agreed at a general level on a philosophy of enforcement, choice of strategies gave rise to many different opinions. EPA officials publicly aired conflicting views. For a period, it was not clear who accurately represented the EPA's position. Officials retracted policies or embarassingly apologized for them. At the same time, the administration replaced a large percentage of the EPA staff; and the Department of Justice, responsible for pursuit of some of the EPA's cases, itself experienced a change of leadership.

Given these organizational obstacles to consistent policy, it is not surprising that some of the most effective environmental programs are administered by counties and state agencies with few employees. The Los Angeles Hazardous Waste Strike Force is an example. A small group of inspectors and lawyers have, through well-publicized, concerted efforts, achieved great visibility in the industrial community. Part of the unit's efficiency derives from members' direct access to the aggressive messages of the head of the strike force. At the state level, efforts of a small staff explain Oregon's achievements in getting compliance with state land use goals. A volunteer support group has also been instrumental in the successes of the land use board.

Successive sessions of Congress themselves lack coordination; statutory change is predictable. Whereas since 1970 Congress has been generally environmentalist in orientation, Congress watchers have identified important shifts in the priority placed on pollution control legislation. For example, in 1981 Congress

---

[25] *Inside EPA*, May 18, 1984, at 11.
[26] *Inside EPA*, Nov. 5, 1982, at 1, 7.
[27] See, for example, "EPA Overcomes Budget Office, Issues Tougher Exhaust Rules," *Los Angeles Times*, Mar. 9, 1985, at 12, col. 1.

passed the Steel Industry Compliance Extension Act.[28] The statute amends the Clean Air Act to allow steel companies to postpone compliance until three years after the originally imposed deadlines. Congress also extended deadlines for compliance with mobile source standards. The business sector readily recognizes its ability to obtain *ad hoc* exemptions to compliance schedules. Serious rationales may exist for postponements in one industry; yet the larger compliance message is that deadlines are movable and that Congress is often willing to make requested changes.

Several characteristics of Congress and state legislatures explain changing environmental objectives: (1) representation changes over short periods—in the United States House of Representatives two-year periods sometimes see major shifts in party composition; (2) legislators recognize the need to modify positions during campaign periods; and (3) oversight activities identify problems of legislation. Affected interest groups can tell convincing stories of law's negative or regressive results even if they are not substantiated and are not representative of overall impacts.

If pursued absolutely, however, consistency in articulation of environmental law can be a weakness of regulatory process and can inhibit general compliance. One commentator has noted that the California Air Resources Board "appears determined not to allow any major philosophical changes in the battle against smog, apparently fearing that any ideological change would open the door to significant weakening of its tactics and authority."[29] When a government agency promotes consistency to prop up its credibility or to avoid controversial judgments and ignores opportunities for flexible interpretation, it may threaten the regulatory policy in the long run.

Bardach and Kagan (1982) have fully analyzed this classical dilemma of rule orientation versus discretion. Several forces move regulators toward consistency in rule articulation and enforcement that may produce rigidity rather than a helpful consistency. Enforcers seek unambiguous rules to ease problems of proof; inspectors welcome tightness of rules because it allows them to assume an air of neutrality in interactions with the firm—blaming regulatory enforcement on superiors. The safest course for regulators is "to go by the book" (p. 301).

If properly directed, some flexibility in regulatory matters can promote compliance. Bardach and Kagan make the case strongly that "reasonableness in not insisting on strict enforcement (helps) elicit a cooperative posture from the regulated enterprise" (p. 138) and offers dividends in effectiveness, whereas

rigidly following the rules leads to unreasonableness and unjust results, in turn pro-

---

[28] P.L. 9723, 42 U.S.C. § 7413(e)(1).
[29] Elias, "Smog Triggers Hobson's Choice," *Orange County Daily Pilot* (California).

ducing legal resistance, greater inducements to corruption and, frequently, surreptitious returns to discretion by agency officials. (p. 153)

The authors plead for increased discretion; they recognize that some political analysts fear that flexibility leads to unbridled authority and possible regulatory capture. But their treatment of potential capture problems is comprehensive, and they conclude unabashedly:

Enforcement officials must be given broadly worded grants of discretion that will allow them to order regulated enterprises to do whatever seems necessary and prudent under the particular circumstances, as well as discretion to relax the rules and tailor their enforcement procedures to the situation. (p. 34)

Achieving the proper balance between consistency in rule articulation and enforcement on the one hand and flexibility on the other is no mean task. Lawmakers are reticent to confer discretion on agencies, especially in health and safety laws sought by vigilant interest groups concerned about excessive influence of the regulated party.

Several conditions can promote reasonableness without sacrificing agency responsibility. Discretion is least risky (1) when agencies have staffs noted for their professionalism and whose tenure is relatively secure; (2) when revolving-door personnel movement between government and the regulated firm is infrequent; (3) when relationships between regulators and businesses are continuous; and (4) when the most efficient means for reaching environmental objectives have not been established and thus flexibility promotes learning within the agency.

In addition, Bardach and Kagan (1982) suggest specific strategies to promote a proper balance between the rule of law and discretion. Agency administrators can provide to personnel the overriding philosophy for reconciling an organization's primary mission with competing social values. Furthermore, administrators should offer ''regular mechanisms for enforcement officials to discuss hard cases'' and training to build competence within the ranks (p. 183).

Without such constraints, flexibility can be a subterfuge allowing government to neglect the difficult administrative decisions necessary to get compliance. A testing ground for the tension between discretion and rule of law is the EPA's ''reasonable efforts program'' which was designed during the Reagan administration as an avenue to compliance for air quality districts which have no chance of being in attainment by deadlines set in the Clean Air Act. The program, nowhere authorized by statute, aims to achieve emission reductions in areas with intractable problems and to avoid imposition of sanctions such as construction bans. Under the program, EPA reviews with nonattainment-districts control measures described in air quality plans and undertakes audits of implementation. The agency cooperatively works with local districts to determine areas for improvement of implementation and to select control measures proven effective in other regions. Advocates hail the plan as a flexible, realistic response to an immensely complicated task, but opponents conclude that it adds another review procedure to a situation that demands implementation of existing plans and enforcement when control measures are ignored.

Rationality of Regulation: Is the Law "A Ass"?

Rationality of regulations also influences business compliance (Bardach & Kagan, 1982; Levin, 1977; Rodgers, 1973, Sabatier, 1977, Skolnick, 1968). No objective standard defines rational law, but some areas of environmental regulation invite indicting criticism. Irrationality covers excessive costs, absence of a regulator's understanding of the nature of the regulated business, unwise priority setting regarding the use of government resources, and inappropriate choice of strategies for meeting society's environmental goals.

To be sure, some charges of irrationality communicate little more than industry's desire not to be controlled; complaints may offer only the regulatee's self-interested perspective. We can distinguish these charges from attacks on government for failing to help business to understand the reasons for and the basis of regulation. Automobile executives continually argue that the Clean Air Act fails to incorporate needed cost effectiveness and cost–benefit criteria. Detroit interviewees contend that the EPA is forcing a "two hundred thousand dollar car" through requirements such as durability tests and high altitude emission control standards[30] and that company employees do regulatory analyses at the expense of producing new and better products.[31] Large industry complains that regulators unfairly concentrate on it because government perceives the availability in the private sector of immense resources. Small business reports that environmental regulation has been applied inequitably, in light of the responsibility of major sources for environmental degradation. Some of these criticisms are no more than editorials and should not be the basis of environmental policymaking. It is as natural for industry to fight for regulations that will protect it as it is "for A&P to sell groceries."[32] Other criticism demands serious response.

Legislative Irrationality: Clean Air, Dirty Coal

And you'll never tell me . . . this process of running around the hall in and out of a conference committee at 11 o'clock at night deciding whether it should be .41 or this

---

[30] In an interview for this study (D) these were described as a one billion dollar cost to consumers that affected only 4% of all the vehicles in the country. (The high altitude standards applied to automobiles that are primarily to be driven in areas where atmospheric conditions dictate special control mechanisms to maintain emission standards.)

[31] Interview D.

[32] See Stigler (1971).

or that or the other thing is a rational process. The people bartering on what the emission levels should be on automobiles wouldn't know a hydrocarbon if they tripped over it. . . . But there they are: "I'll give you this, if you give me that." It's almost like you're out in Nevada.[33]

This auto company executive's description of the passage of environmental law introduces a significant attack on regulatory rationality. There are several focuses of this criticism of environmental lawmaking: unlikely interest group coalitions opportunistically promote different objectives in a single regulatory package; the approach to control is piecemeal; and the information base that is the background for rule promulgation is limited.

Ackerman and Hassler (1981) describe the strange coalition that formed to promote an air quality regulation requiring that sulphur scrubbers be added to tall stacks on all facilities burning coal, even those that burn low sulphur coal. Environmentalists and representatives of the coal industry on the East Coast lobbied effectively together for the control strategy, clearly for different reasons. Business-oriented and other media criticize such uniform approaches to regional air pollution problems. "Off the shelf" regulations require overcompliance by some businesses in order to control a particular industry sector sufficiently.

Proponents of environmental regulation may also include those with financial interests in the proliferation of rules, no matter what the substance of such rules. During the early Reagan administration's efforts at deregulation, the negative impact of decontrol on the demand for legal services became apparent. Observers concluded that lawyers based support of a large federal regulatory program more on self-preservation than on their stated reason: to avoid proliferation of regulation among the fifty states. One Washington attorney noted, "We had oil price regulations all over the place, and it created an absolute lawyers' paradise but never protected anybody against anything. . . . It was ludicrous."[34] *The New York Times* reported that the lawyers were "suddenly supporting their erstwhile foes."[35]

No doubt, resulting environmental laws are on occasion of questionable rationality. A Brookings Institution study found the Clean Air Act to be structurally deficient. Brookings attributed much of the progress in air pollution control in the 1970s to national economic downturns and shifts away from use of coal and other polluting fuels—shifts dictated by factors other than law. The study recommended increased discretion and authority for the EPA administrator, whose actions in implementing the act should be guided by cost effectiveness and risk analyses. Critics cite numerous other regulations as seeking noble goals but being unreasonable. These include Federal Water Pollution Con-

---

[33] Interview C.
[34] Jack A. Blum of Blum and Nash quoted in *The New York Times,* June 1, 1981, at 21, col. 3.
[35] *The New York Times,* June 1, 1981, at 21, col. 3.

trol Amendments calling for zero discharge into the lakes and streams of the country by 1985 and the reporting provisions under the Toxic Substances Control Act which require premanufacture notification to the EPA for new chemicals. Implementation of TSCA has created a paper deluge at the agency. The potential workload has required the EPA to review notices selectively. Occupational health rules are even more open to ridicule and include OSHA's notorious control of toilet seats (they must be split and not round), its stipulation of the heights of ladders that can be used on the job, and its prohibition on ice in drinking water.[36] Bardach and Kagan (1982) explain that regulatory unreasonableness results from the unwillingness of functionaries to make substantive decisions about the effects of rules, in part because of immense uncertainty in their information base. Regulators like to apply unqualified legislative statements rigidly, such as the Delaney amendment in food and drug law which prohibits the sale of food additives and products which are found to cause cancer in laboratory animals. Such a posture removes agency personnel from scientific controversies that plague rule making (discussed in the next section).

Other attacks on the rationality of regulation are more global. Industries affected by single-medium legislation (the Clean Air Act, the Water Pollution Control Amendments) point to government failure to devise overall strategies for reaching national environmental objectives. Confused businesses cite land disposal problems arising from prohibitions on water disposal. By concentrating on a single medium, environmental control strategies continually shift the focus of the problem but do not decide ultimately what is to be done with residuals of production processes.[37]

These attacks on environmental regulation are based on a high standard for government rationality. Whatever the merits of criticisms, industry's point can often be made convincingly with the result that citizens see the effects of poor environmental regulation more clearly than failures of other regulatory policies. Business can easily communicate the alleged irrationality of rules to the layperson when the subject matter is one of everyday experience. The nonexpert does not as readily understand regulation of other activity, such as banking and monetary controls.

The sheer number of health and safety regulations also erodes the respect accorded individual environmental rules. Proliferation of law tells business that the analysis undertaken for each provision and the import of each regulation are limited. When regulatory schedules take up volumes of state and federal codes and deadlines for promulgation continually are reset, the regulated readily ques-

---

[36] This rule was developed when ice came from rivers, some of which were polluted (Clay, 1983).

[37] In 1982 the EPA again raised the possibility that environmental regulations should be integrated across all media. A policy analysis indicated that a single open ''organic act'' might be an improvement.

tion the quality of standards. The volume of rules was a major factor in the passage of a radical amendment to the California Administrative Procedure Act.[38] The amendment created the Office of Administrative Law which reviews and has authority to disapprove both existing and proposed regulations. The California legislature statutorily established standards for evaluation of rules; they are necessity, authority, consistency, clarity, and reference in law. The objective of the amendment was "to reduce the number of regulations and to improve the quality of those regulations which are adopted."[39]

The Scientific Basis of Regulation

> I've been told TCE (Trichloroethylene) is a suspected carcinogen and I'd like to know if it does cause cancer, or if the body can throw it off.
>
> —Navy veteran[40]

> The essentials of this situation will be understood by many Americans. . . . Scientists who hid nature . . . are not to be trusted. A bomb has been dropped on the Carsonian religious-political parable which is the only meaning Americans have ever been given for "environmental cancer."
> When that parable explodes, as it must, and the public understands the magnitude of the ideological delusion in which the entire country was enmeshed, a few other, and more sophisticated, questions will be asked. For example: While the Biologist State was concocting a pseudo-science and regulating industry on the basis of a fairy tale, while it was manipulating theory and data the way a cardsharp shuffles cards, while it was suffocating American minds with myth . . . *where were the critical scientists who knew that this was happening?*
>
> —Edith Efron in *The Apocalyptics* (1984, p. 423)

Perhaps the most frequent criticism of the rationality of environmental regulation concerns the science that forms the background for standard setting. Challenges to the information base of environmental controls are made in several forms: Data are lacking, biased, incomplete, or irrelevant. Adversaries conclude that scientists "on the other side" deliberately distort the meaning of findings and results. Rather than being a foundation for the rational development of generally accepted rules and regulations, science is often the source of charges of government irrationality in environmental and occupational matters (Brooks, 1975).

Health effects research is an enlightening case in point. Controversies over the scientific basis for determining health effects of pollution are numerous. Government, industry, and environmental groups often employ science differ-

---

[38] A B 1111, Ch. 567, Stats. 1979.

[39] Cal. Govt. C. § 11340. For a fuller presentation of the amendment see *The California Regulatory Law Reporter* 1 (Spring 1981).

[40] *Los Angeles Times,* Sept. 24, 1979, Part 1, at 3C.

ently in addressing the advisability and legality of environmental rules. Regulator and regulatee will differ on which information is relevant and probative. In a pluralistic society such differences are predictable, but they raise serious questions about the rational basis of standards. Critics address uncertainty and ambiguity in scientific conclusions and value differences that fill in the informational gaps. The debate is particularly intense over health effects because environmental controls are immediately visible and expensive whereas pollution's health effects may be of long term and causal links are difficult to establish and virtually incomprehensible to the lay public.

Environmental health regulation raises questions about how to resolve disagreements among scientists and other experts when there is uncertainty about cause and effect, about how to address short-term regulatory needs when scientific answers will not be available for many years, and about how to regulate when scientific capacity to detect pollution outstrips capacity to determine whether the pollution is serious.

Uncertainty is the norm in environmental health research. Relating insult to human population effects is immensely complex. For example, despite a mammoth scientific effort to discover causes of cancer, only twenty-six of the hundreds of thousands of commercially used chemicals have been shown to have carcinogenic effects in humans. But public concern over the effects of chemicals is great; and that concern translates to fears about carcinogenicity, mutagenicity, tetragenicity, and other outcomes of use of the industrial products. Apprehension translates to calls for regulation. The adversarial nature of scientific contributions to regulation, whatever its merits, has weakened the reputation of the rational basis for rule promulgation. In almost every major environmental controversy in the last two decades, scientists have confronted other scientists on the effects of proposed actions. Acid rain is among the most famous recent cases. Environmentalists have identified effects of acidification which demand immediate controls. EDF concluded that midwestern power plants are probably responsible for the acid pollution of lakes in the Adirondack Mountains in New York State. At the same time, a geologist hired by Con Edison testified that naturally acidic materials in the area around the lakes are being carried by rain and snow into the waters;[41] and the electric power industry argued that lake acidity is affected by other factors as well, including local land use patterns and vegetation cover. Environmentalists, responding to a study that found only 75 of 790 Adirondack lakes totally fishless,[42] asserts that the number of fishless lakes in a region should not be the indicator of the acid rain problem. Thus the debate enters another phase.

Scientists do not disagree on all parts of the analysis. Acidification does reduce bacterial decomposition of organic matter; plants tolerant of low pH will

[41] *The New York Times,* Mar. 17, 1983 at B2.
[42] *Inside EPA,* June 1, 1984, at 2.

develop in acidic bodies of water and these can choke out other species; fish populations are depleted or eliminated. But causes and extrapolation to human health effects remain questions of major disagreement. Air flow patterns, natural sulfur emissions levels, the life time of dry deposited SO_2, and interactions among contributors to acid precipitation are variables in the causal equation that provide considerable room for scientific dispute. Such differences also arise in the cases of dioxen, formaldehyde, and some of the criteria pollutants.[43]

Even in the Love Canal incident, where everyday knowledge held that living near the abandoned dump site must be harmful, scientific studies reached different conclusions about the link between exposure to chemicals and chromosomal damage. Additional studies are underway. As a complicating factor, the government investigation discovered levels of dioxin in the Love Canal area higher than originally measured. Objective analysis concluded that those living in the contaminated area had a significant increase in spontaneous abortions but did not experience increased incidence of anemia, chloracne, asthmatic conditions, convulsive disorders, or congenital defects.[44]

Two cases illustrate the particular nature of scientific controversies.

Malathion, the Med Fly, and Health Effects. California Governor Jerry Brown's political problems in his second term arose in part because of the absence of unambiguous scientific information about the impact of malathion on human health.

In late 1980 the Mediterranean fruit fly problem in California became extremely serious. The med fly reproduces vigorously, and it posed a major threat to the huge California agricultural industry when the insect was discovered in northern and central areas of the state. The governor initially favored eradication of the pest by nonchemical means and certain well-tested pesticides. An executive order required homeowners in affected regions to strip their trees and dispose of the fruit in prescribed ways. Results were limited.

The major competing approach relied on use of malathion, a pesticide in domestic use for thirty years and a chemical employed to control, among other things, children's head lice.[45] Whether malathion, which had been used in Florida and Texas to rid agricultural areas of the med fly, has significant human health effects became the central question in the decision on controls. The governor entered the debate confronted with a number of different conclusions on the health effects of the substance but with no good studies on the long range impacts.

The State Health Department reported to the governor that the health effects

43 "Dioxins Health Effects Remain Puzzling," *Science* 221 (Sept. 16, 1983); Crandall and Lave (1981); and *Proceedings of the Consensus Workshop on Formaldehyde*, Little Rock, Arkansas, Oct. 4–6, 1983.

44 J. P. Murray, J. J. Harrington, and R. Wilson, "Chemical and Nuclear Waste Disposal," *Cato Journal* 2 (1982), 565–606, at 592–593.

45 *Los Angeles Times*, Feb. 26, 1984 at 1, col. 3.

of malathion spraying were "insignificant," and studies of both the National Cancer Institute and the National Centers for Disease Control indicated no relationship between spraying of malathion and illness. A University of California entomologist also advised Governor Brown. The scientist's position reportedly changed at least three times during the period of the crisis, but overall he concluded that the health effects were significant. Other experts lined up on each side of the controversy. One pathologist cited a danger but refused to submit his work to peer review.[46] Governor Brown later described a situation that faces many policy makers:

> Everyone says [the expert] doesn't have any proof and he says malathion will mix with tissue and it has a significant possibility of long-term mutagenic effects; that's what [he] believes and he's a respected pharmacologist.
>
> It was very hard for me to sit there and judge with the time pressure on me, to become a judge in a scientific debate.[47]

Donald Kennedy, President of Stanford University and formerly Director of the United States Food and Drug Administration, later summarized the health effects controversy: "There were a lot of uninformed or misinformed people . . . to the point of being irrational on the issues . . . a lot of strong feelings had been stirred in opposition."[48] Under such conditions, whatever controls regulators adopt are certain to alienate sections of the business community.

When Rats Die and No One Knows Why. Whether formaldehyde should be treated as a carcinogen or regulated on the basis of other suspected health effects has been a persistent question in occupational and environmental regulation. Formaldehyde is a commonly used substance found in 12,000 different products; 8.6 billion pounds of the material are produced annually. The formaldehyde industry and associated industries are valued at approximately $235 billion a year; the public policy decisions on its control are both socially and economically significant and thus controversial.

The disagreement over formaldehyde's effects is gleaned from the divergent ways in which agencies have addressed its regulation. In 1980, the Consumer Products Safety Commission initiated a ban on the installation of urea-formaldehyde foam; but, at about the same time, HUD chose not to regulate formaldehyde in mobile homes. In 1983, a federal court of appeals overturned the Consumer Products Safety Commission's ban. The EPA position during this period was mixed, but it decided not to regulate.

Then in May, 1984, amid charges that the EPA had deviated from its own cancer risk assessment policies in bowing to the formaldehyde industry, the EPA designated the substance for regulation under the Toxic Substances Control Act, Section 4F. The EPA's working assumption was that the cancer risk from ex-

---

[46] *Los Angeles Times*, Feb. 26, 1984, at 1, col. 2.
[47] *Los Angeles Times*, Aug. 3, 1981, at 16, col. 1.
[48] *Los Angeles Times*, Aug. 3, 1981, at 3, col. 5.

posure to low doses of formaldehyde, which was based on a conservative linear projection, would be about 1 in 10,000 or between 4 or 5 million people—primarily workers in the clothing industry or residents of mobile homes. Because it was classified as only a weak carcinogen but was widespread, it was decreed that formaldehyde should be considered a substance posing a substantial risk.

The EPA's decision was not greeted with unanimous favor even at the EPA. Several observers within the agency had hoped to wait for results of a soon to be released massive epidemiological study on formaldehyde effects on apparel workers. The Office of Science Technology and Policy (OSTP) and OMB continued to question whether there was enough animal evidence to regulate the substance; both agencies concluded that ''the animal evidence shows the cancer effect of formaldehyde is not linearly related to doses at low levels, suggesting existence of a threshold, and thus EPA's use of a conservative, linear low-extrapolation model'' was invalid.[49] And critics charged that the EPA had ignored the conclusions of a 1983 consensus workshop on formaldehyde. EPA officials countered that the decision was compatible with the workshop's conclusions.

Observers, no matter what their views on the wisdom of regulation, desire greater certainty about the effects of formaldehyde than the research allows. A representative of the Consumer Federation of America complained that ''the problem is that no one has set a standard for exposure to the gas. We're trying to get the government to say if the level is below a certain amount there shouldn't be any problem''[50]; *The Wall Street Journal,* on the other hand, in an editorial titled ''The Rat That Died,''[51] concluded that the risks to humans of formaldehyde were insignificant. The editorial argued that the rat studies of the health effects of the substance were not generalizable to other species because rats have a unique breathing system which makes them particularly vulnerable to exposure to formaldehyde. The editorial advocated nonregulation in light of immense costs associated with controls.

Considerable scientific disagreement over the impact of formaldehyde exists despite the fact that more than 700 scientific papers have been published on its health effects and hundreds of scientists are working on the issue of control. Scientists and policy makers even disagree on what formaldehyde is and which substances should be regulated. Some government officials wish to see regulation limited to formaldehyde resins used in making apparel, while others would regulate treatment of wood products in mobile homes, some conventional residential uses, and a wide range of cosmetics and other products. About eight hundred cosmetic products contain formaldehyde.

[49] *Inside EPA,* May 25, 1984.
[50] *The Wall Street Journal,* Dec. 14, 1983, at A-1, col. 6.
[51] *The Wall Street Journal,* Sept. 19, 1984.

The complex environment for regulating formaldehyde derives from the absence of scientific consensus on several issues. In epidemiology scientists differ on whether observed cases of health effects should be compared only with local populations or with the national population; what classifications of exposures should be used (should high exposure groups be subdivided into more discrete classes?); how to combine data in different studies to undertake an overall risk assessment; and how to treat latency (some epidemiological studies show a small number of cases of cancer and other health effects at the time of publication, but the numbers increase as the study group is followed after publication of the initial research report).

Other phenomena in the study of cancer epidemiology of formaldehyde add to the regulatory controversy. These include the autopsy effect (some groups are more likely to be autopsied then others); behavioral confounding factors (members of groups, such as embalmers, that have been highly exposed to formaldehyde may be at higher risk because of other health-threatening behaviors including alcohol consumption); and end point disagreement (many studies lack scientific rigor in determining whether a health effect was actually experienced). What is more, the very definition of an end point is of concern. For example, few scientific observers classify dermatitis as significant; yet cancer cannot be the only health effect of concern to regulators. Overall, both the development of epidemiological studies and reporting of results have lacked standardization.

Epidemoiology is not the only source of scientific disagreement. Those who study carcinogens differ as to the significance of benign tumors and preneoplastic lesions associated with exposures, as to whether formaldehyde is an initiator or promoter of carcinogenesis, and whether the identification of tumors related to exposure in rats and not in mice should be a source of concern or of relief for students of human health.

Even minor differences in data interpretation can lead to very different regulatory conclusions. The data base available to pathologists in one major study was so small that investigators disagreed on whether the number of malignant tumors, out of a universe of six under investigation, was two, three, or four. A difference in classification of even one tissue significantly affects a risk assessment.

Other aspects of formaldehyde research have similar gaps. Histopathologists are unclear as to formaldehyde flow patterns in organisms and interactions with other materials within the subject. Consequently, it is not known whether formaldehyde is the cause of distal tumors, that is, those physically removed from the nose, the main source of entry of formaldehyde. In immunology, researchers wonder whether formaldehyde elicits a response only in desensitized populations or is a primary sensitizing agent. Environmental scientists admit that there is an absence of standardization in research protocols (e.g., should windows be opened or closed in houses in which formaldehyde exposures are being studied?) Monitoring is complex, and it is not clear how to report peak exposures of formaldehyde because sample taking does not always control for temporal and spatial variability in research settings. To compound the ambiguity, formaldehyde is used in treating certain health problems, such as urinary infections; and in some countries it is employed as a cleaning agent.

Thus, those who do risk assessments of formaldehyde as a basis for regulatory policy face very different scientific understandings. Scientists bring to the study of formaldehyde, as to many potentially suspect substances, different exercises of scientific judgment when insufficient good data exist. Finally, the choice of scientific techniques for presenting data also varies. During the risk assessment panel meetings at the 1983 consensus workshop on formaldehyde, scientists argued over (1) which mathematical model to employ in fitting the small number of unambiguous data points, (2) whether to use data on rats exclusively or to include data on mice, (3) whether to add a safety factor to risk assessments even if an observed "no effect level of formaldehyde exposure" could be found, and (4) whether to extrapolate routinely across species in the absence of clear scientific reasons to do so.

In short, policymakers must address the control of a widely used substance about which we have considerable scientific information and massive scientific disagreement. Fodder for charges of irrationality is everywhere.

The Gaps in Science. Cases like those above illustrate the bases for charges of irrationality in environmental policy making. For policy and political reasons, government must regulate with insufficient information. Several questions which must be answered to convince industry are transscientific; that is, they are not subject to resolution through the normal procedures of science (McGarity, 1979). Some examples: animal experiments to test a suspected substance at an ambient level may be prohibitively costly. As in the classic "mega mouse" case (employing the number of mice necessary to reach statistically significant conclusions), the test may be logistically impossible. As we have seen, extrapolation of animal results to humans is not always convincing, and extrapolation of high-dose laboratory data to low-dose ambient levels can employ many different mathematical assumptions. Resulting predictions of incidence of health effects differ by orders of magnitude. A related issue concerns the interpretation of data generated from high doses over short periods of time to comparatively low doses over a lifetime.

When answers are practically achievable through scientific means, much work remains before scientists and regulators reach consensus. For many health problems, theories of causation are multiple; only a few involve environmental contamination. Fundamentally, researchers cannot always identify suspected contaminants and model streams, air, water, land, and routes within these media, in which the pollutant travels. We have only elementary knowledge about interactions with other forms of the suspect substance and with other substances in the environment, about pathways in the body, and about biochemical reactions in various routes of entry. Within the body there are numerous potential combinations with other toxic or hazardous materials, and with naturally occurring substances that are benign at low levels. Cigarette smoking and other social behaviors confound analyses of effects. Finally, cultural, psychological, and individual differences in response to disease complicate the environmental sciences. Such uncertainties lead to debates about the very need to regulate and about the specific rules once regulators decide to impose restrictions.

Industry finds standard setting under conditions of great uncertainty particularly disturbing. Since the science of detection is outpacing the science of causation, industry representatives fear the prospect of control to parts per trillion of emissions. If agencies regulate to the point at which pollution is identifiable, no limit exists to controls. When science is unable to give unambiguous results or when science cannot give clear direction, a regulatory dilemma arises; the issue, crudely put, is whether one waits for a dead body count before one regulates. So stated, environmentalists beg the question, for some potentially regulatable entities probably will never demonstrate ill effects. Time is not the issue; rather, causation is. Science can now only speculate on actual outcomes.

This fundamental difference in use of science drives many discrepancies in views about rule rationality. The EPA based consideration of rules regulating gasoline storage tanks, with costs to industry that ranged up to $4.7 billion, on "skimpy and anecdotal information on the health and environment effects" associated with leaking underground tanks.[52]

Two recent California examples are also relevant. In a case involving the Aerojet Company, test results from the California Department of Health Services indicated that company wells had levels of trichloroethylene (TCE) ranging from 4.3 to 470 parts per billion. Health department officials concluded that, although no standards exist for TCE in drinking water, any contamination should be considered harmful. Aerojet officials strongly opposed the suggested controls.

Health effects of vinyl chloride exposure were at issue in cases involving chemical and rubber plants in the South Coast Air Quality Management District. Research indicates that vinyl chloride can cause liver and brain cancer, blood cell damage, miscarriage, and birth defects. Industry representatives assert it would require massive exposures many times the levels emitted at the California plants to affect health. A California deputy attorney general who represents the State's Air Resources Board explained why the state set a stringent standard for the substance:

[52] *Inside EPA,* June 8, 1984, at 9.

"We obviously can't wait until 20 or 30 years to see if people develop angiosarcoma" (a type of liver cancer associated with vinyl chloride exposure).[53]

A 1981 EPA study concluded that latent health problems may be linked to exposure to vinyl chloride. California determined that no exposure to the material could be presumed safe; the state set the standard at the lowest detectable level. One of the affected companies sued to invalidate the rule, contending that there is no proven relationship between concentrations of the chemical and any adverse health reaction.

The better-prepared industry is aware of problems in the informational basis for regulation. The EPA Community Health and Environmental Surveillance System ("CHESS") program is a notorious case in the history of the scientific bases of regulation. CHESS was developed in 1970 as a series of epidemiological studies in several communities to determine the effects on humans of sulphur dioxide, other oxides of sulphur, and other air pollutants. CHESS's fundamental objectives were sound: to evaluate existing environmental standards, obtain health intelligence for new standards, and document the health benefit of air pollution control.[54] However, assessments of the program paint a picture of a research project plagued by EPA-imposed time constraints, budget limitations, and technical incompetence.[55] The project produced some results which were scientifically worthless. CHESS failed to undertake normal quality control, to validate methods, to cross-check data, and to calibrate instruments. Certain results cannot be generalized to communities other than those studied. A congressional investigatory committee, the Brown Committee, commented:

> The overall large error band on all measurements (possibly exceeding 100%) and the apparent bias of SO_2 concentration data on the low side, make most of the numbers presented in the CHESS Monograph unusable.
>
> EPA should not use data in the 1974 CHESS Monograph as supportive justification for policy decisions without explicit qualification and should avoid unqualified reference to the document in public information statements or documents.[56]

Industry representatives became cognizant of weaknesses; they described CHESS as based on unverifiable statistical analysis performed by the EPA, not on "professional, technical standards":[57]

> The problem is that a lot of this stuff got lodged in citable technical literature, and the Brown Committee of Congress investigated it and basically crapped all over it, and yet it still crept into the new criteria. . . . Some other renowned statistical people repli-

---

[53] *Los Angeles Times,* Dec. 14, 1981, Part 2, at 2, col. 1–2.

[54] U.S. Environmental Protection Agency (May 1974, at 1–4).

[55] For example, see Report, 1976.

[56] Report, 1976, p. 2.

[57] Interview B.

cated all the studies; and, absolutely, there's only one of them that was admissible by
any professional, technical standards. Yet it still showed up in the draft because EPA
was writing the draft.[58]

The CHESS fiasco did immeasurable damage to EPA's scientific reputation; and CHESS's termination left the United States with no means of scientifically measuring the health benefits of air pollution control.[59]
Industry is also sceptical about the scientific basis for regulations on a more
global level. Edith Efron in *The Apocalyptics* argues that the federal government
became captured by an ideological conclusion that chemicals in the environment
were a main cause of a dramatic increase in cancers experienced in the United
States from 1969 to 1976. Building on an environmental position which began
with Rachel Carson's classic *Silent Spring,* Efron maintains, several government
agencies unscientifically concluded that a cancer epidemic existed and that it was
linked to industrial activity. Two British scientists, Sir Richard Doll and Richard
Peto, later refuted this assessment. Doll and Peto determined that only about 8%
of deaths caused by cancer in the United States can be attributed to industrial
carcinogens (Nisbet, 1984). Opponents of occupational and environmental regulation often cite this reaction by American government scientists to a phenomenon that might be based on the aging of the American population.
These cases exemplify what industry concludes is a rush to judgment fired
by a desire to change fundamentally the country's orientation toward growth and
economic development.
Scepticism and use of "science as weaponry"[60] even counter agreements
on the most promising approaches to increasing scientific certainty in rule making. Each scientific strategy has its weaknesses. The regulatee may see quantitative risk assessments as hocus-pocus or government sleight-of-hand because
of nonstandardized choices of mathematical models or of health "end points."
As the cases in this chapter note, epidemiological studies are hampered by
mobility of the population studied, lack of standardization in the discipline,
variable characteristics of the areas surrounding the studied group (such as indus-
trialization), diet and behavioral patterns, socioeconomic and age differences,
long latency periods of effects, disagreement on comparison groups to describe
expected incidences, and other factors. Some of these items are extremely difficult to control. Generalizing from whole animal studies and from short-term
laboratory assays to human health effects is also filled with problems.[61]
Different opinions on how to resolve these uncertainties explain variations

[58] Interview B.
[59] Comments by James Whittenberger, Director, Southern Occupational Health Center, University
of California, Irvine, January 24, 1985.
[60] Marc Roberts, Stephen Thomas, and Michael Dowling use this expression (1984).
[61] See discussion above.

in federal environmental health policy, such as decision rules for concluding that a substance is cancer-causing.[62] There appear to be Republican and Democratic positions on whether (1) benign tumors are significant in cancer risk assessment, (2) interspecies extrapolation of data and results of studies is appropriate, (3) high to low dose extrapolation makes scientific sense, (4) extrapolation across species requires comparability of exposure routes,[63] and (5) epidemiological data are better evidence of carcinogenicity than short-term tests and as valuable as long-term animal tests. Almost all studies trade off one problem of methodology for another. Adversaries in the regulatory process will easily find the Achilles heel of an investigation in order to undermine its applicability or, at the very least, to postpone the date of a policy.

Some issues related to the rationality of regulations are value questions hidden in scientific debate. Which effects should be considered problematic? Is long-term morbidity the ultimate outcome of concern? What is a true health effect? What about psychological impact of environmental stressors? A United States Court of Appeals required analysis of psychological impacts in environmental impact reporting; the case involved the decision on whether to reopen the Three Mile Island nuclear power plant. The court concluded that the Nuclear Regulatory Commission had ignored "the simple fact that effects on psychological health are effects on the health of human beings," thus differentiating low-level socioeconomic anxieties from true stress. The United States Supreme Court later rejected the position but not before it had aroused considerable industry concern over how suspected correlates of psychological stress might be regulated.

In the most egregious cases, allegedly scientific analysis hides policy conclusions. In 1982 it became known that the proposed standards for dioxin were being influenced by the aggressive involvement in EPA decision making of a consultant "science advisor." The advisor performed a risk assessment and concluded that exposures which were orders of magnitude above the existing standard would be acceptable. EPA internal correspondence described the involvement of the advisor as "forceful" and expressed fear that if the advice were taken the publicity effect for the agency would be "ghastly."[64] The memo went on to state:

> The Superfund people, the office of waste programs enforcement and my office are concerned, however, that because [the advisor] is aggressive and appears to have the [assistant administrator's] ear, he may get the chlorinated dioxin work group's recommendations reversed (he has been extremely adamant at meetings). Because the Su-

---

[62] See "Revisions in Cancer Policy," *Science* 36, Apr. 1, 1983.
[63] See Interagency Risk Management Council, "Assessment of Cancer Risk" reproduced in *Inside EPA*, Apr. 13, 1984 at 14–15.
[64] *Inside EPA*, Nov. 19, 1982, at 5.

perfund office and the office of waste program work for [the assistant administrator]
they are reluctant to take on [the advisor].

The Role of Science: Conclusions. Uncertainties, ambiguities, and opportunities for real and feigned disagreement in the scientific enterprise suggest the
extent to which conclusions about the rationality of regulation are inherent in the
regulatory process. Looking to science for objective referrents to acceptable
controls can assist at the margins, and in some special cases, but scientific
conclusions will not generally satisfy those affected by expensive regulations.
Indeed, science will be used to generate alternative understandings of what is
rational.

Some commentators even argue that scientific interpretation is a process
inherently affected by culture and that little hope exists for an overall consensus
on the health effects of pollutants. Levine (1982) concluded that the social
locations of the agency officials and scientists in the Love Canal incident determined interpretations of scientific facts. Career and organizational demands
drove the choice of research strategies in part because scientists and policymakers generally refrain from generating findings with unpleasant or detrimental
social or economic consequences.[65]

Public perceptions of the rationality of regulation affect a business firm's
response to rules in another way which we address in the next chapter. Common
interpretations of evidence that links environmental conditions to health effects
and to environmental damage can mobilize groups who influence behavior of the
firm. Roberts and Bluhm (1981) describe some of the dynamics in their study of
utilities:

> Changing scientific and technical information helped provide and crystallize public
> reaction in some cases. In Los Angeles, public opposition to additional fossil-fuel
> power plants increased when the key role of NOx in smog conditions was clarified.
> Evidence that the smoke plumes of Ontario Hydro's generating plants were linked to
> ambient SO_2 readings under some conditions likewise encouraged additional public
> action.
>
> Accumulating evidence not only encourages outside activists, but influences
> regulators and those inside the regulated organizations. All parties find it harder to
> defend inaction to themselves, to the courts, and to other agencies when all agree that
> there is a problem. (p. 330)

Commentators suggest numerous institutional reforms to make regulatory
science more convincing and more generally accepted. Ideas range from greater
use of the adversarial process (Majone, 1979) and increased counterbalancing
participation in the record building phases of rule making[66] to the establishment

---

[65] Molotich, H. L. Review of *Science, Politics and People,* 1982, in *Science* 216 (June 1982), at
1401.

[66] The suggestion was made by John Braithwaite. See note 15 *supra.*

of science courts which would separate out and give an objective judgment of scientific fact in complex regulatory cases (Kantrowitz, 1967). Perhaps more reflective of the true challenge to scientific objectivity are less ambitious reforms which seek to gain consensus on issues specific to proposed regulations.[67] Brooks (1984) has recommended using regional research institutions which would adopt findings of pure research to the pecularities of local needs. He points to the apparent success of this approach in agricultural research. This latter group of suggestions will be discussed in Chapter 8.

---

[67] See, for example, the Health Effects Institute and the consensus workshops described in Chap. 8.

CHAPTER 7

THE ACTORS

There was an apparent belief in a great many members of Congress that you can tell the auto industry to do anything; the only reason we were not doing these things is because really we were sons of bitches. In plain language that's what they thought. We wanted to dirty the air; we wanted to kill people in our products; we wanted to use a lot of gasoline because we were in cahoots with the oil companies.[1]

Compliance is a human activity outcome. People perform all the activities and transfer all the messages that create compliance, undercompliance, or over-compliance. With varying degrees of autonomy, people determine those procedures, standard and otherwise, that create or avoid environmental violations. To understand compliance one must understand variation among the individuals and groups involved in the regulatory process. Variations among people who fill roles in government, in the business firm, and in advocacy groups are especially significant when norms are both divergent and in transition. This is the case in environmental law.

Agencies with quite different characteristics formulate regulations and enforcement policies. Rules are directed to firms which vary along many dimensions. Moreover, independent of the quality of law and of the nature of enforcement, government and business actors affect the compliance outcome—whether regulations are strictly followed, agressively attacked, honored in the breach, or virtually ignored. Groups with interests only in environmental quality also influence regulatory interactions.

What Makes a Difference in Public Agencies?

We focus first on government. The reputation of the lawmaking body correlates with the efficacy of rules. Several studies have identified a positive rela-

[1] Interview D.

tionship between high status of the communicator and degree of compliance with a message (Faley & Tedeschi, 1971; Hall, 1977; Tapp & Levine, 1972; Stone, 1978; Wilson & Rachal, 1978). Legislative statements carry more weight than opinions of the judiciary, and these are followed in influence by administrative agency directives (Faley & Tedeschi, 1971; Horai & Tedeschi, 1969). Wilson and Rachal (1978) have noted pointedly that environmental commands to highly paid, highly respected corporate executives are often created, communicated, and enforced by people with lesser status—the proverbial faceless bureaucrats. Roberts and Bluhm (1981) found that the more professional the agency, the more responsive was the firm to its directives:

> The more technically knowledgeable regulators are, the more resources they have, and the easier it is for them procedurally to impose sanctions, the less attractive companies will find it not to comply. Conversely, consider the incentives in a situation like that of Boston Edison, whose air quality violations for years brought routine reprimands; only after heated public protest did the matter wind up in court, where it was settled by a promise to comply. There is little incentive for an organization to be responsive if the weakness of a regulatory agency confronts it with such a situation.

The public-private sector hierarchy is significant in environmental law since, with few exceptions, the source of pollution control regulations is the government bureaucracy. Although legislators and judges play important roles in the regulatory process, the bureaucracy, at least since the New Deal, has carried the largest burden in making law work.

Agency differences that are important in compliance include access to resources, coordination between policy and enforcing units, professionalism of chief administrators, use of resources, and the expertise lodged within critical parts of the agency.

Agency Access to Resources

An obvious factor related to agency influence is the resources it has to pursue compliance. This point need not be belabored, but an example and some qualifications can be offered.

A 1984 report by the State and Territorial Air Pollution Program Administrators concluded that several areas in the country had shifted from attainment to nonattainment status under the Clean Air Act "largely because" of EPA funding cuts. The report, based on a survey covering thirty-four states, concluded that if federal funding were increased by 15%, then 90% of the states would increase activity to control toxic emissions; 50% of the states would increase their activity to determine noncompliance; 42% would step up air quality monitoring; 35% would add to their enforcement activities; and 46% would act further to control total suspended particulates.[2]

2 *Inside EPA*, Apr. 13, 1984, at 2.

The report is notable for its precision, not for its argument that agency access to funding is essential in bringing the corporate sector into compliance. At some level, no one can disagree; resource availability stands at the top of the list of factors related to efficacy of the public sector, although opponents of regulation often paint compliance as intractable whatever resources are applied. Without monitoring personnel and equipment, enforcement specialists, funds for expert witnesses, monies for research of high quality, and competitive salaries, agencies will face major obstacles to reaching their goals.

Nonetheless, as this chapter elaborates, financial resources alone are not sufficient to develop a solid incentive and enforcement record. Indeed, there may well be a point at which additional resources do not generate additional administrative influence. Availability of funds may make it unnecessary to think critically about strategies to pursue, may generate agency growth that is inversely related to internal coordination, and may dull the aggressive posture of a smaller, leaner administrative bureau.

Coordination of the Government Effort

Characteristics of both the agency that articulates rules and its enforcing unit are individually important; so, too, is their interaction. If the relationship between enunciator and enforcer is not well developed, or if communication between them is poor, the influence of mandates is limited. At both the state and federal levels in environmental law interunit relations vary considerably.

The Los Angeles Hazardous Waste Strike Force incorporates a well-coordinated governmental strategy for promoting compliance with environmental rules. The force is composed of representatives from the Los Angeles County Health Department, the city police, the city attorney, the Bureau of Sanitation, and state agencies. Strike force lawyers are trained in both environmental and criminal law. They work effectively with the task force's technical experts, are professionally ambitious, and clearly wish to create reputations of effectiveness for the force.

In several cases, task force members have effectively gathered information about an alleged polluter, carefully communicated this data to those who monitor a pollution source, collected evidence in a "raid" that is admissible in legal proceedings, and aggressively publicized the nature of the violations and the government's capacity to respond. The unit has demonstrated a concern for the timeliness of steps in bringing an action against a polluter. Results include numerous significant victories and, perhaps more importantly, widespread favorable news coverage of the city's power against even well-endowed polluting companies.

Opposite cases are more common. A former chief of the hazardous waste section of the United States Department of Justice offered an example in congressional hearings. He asserted that under a previous administration the EPA had effectively "destroyed the link between EPA and the Justice Department. . . . Justice must start from scratch every time a referral is made. . . . If in the private sector lawyers were called in right before [the client]was ready to go to court, the lawyers would be enraged."[3]

A former assistant United States attorney described his frustration with EPA attitudes on the need to gather evidence for enforcement:

> EPA people . . . don't have the FBI mentality. They are engineers and academics. You need not only agents but a bureaucracy that supports them.
>
> I had fights with EPA over getting samples analyzed. "We have other kinds of priorities," EPA said. If you do get enforcement going, people stand up and listen, but I was up against an agency that was reluctant and inexperienced in enforcement. There must be some changes in the federal government before there will be substantial progress in the environmental field.[4]

Administrator Qualities

Influence of high-level agency personnel on compliance can be immense. The 1983 appointment of William Ruckelshaus as administrator of the EPA may suggest the extent to which observers conclude that an individual can set the direction for even a huge federal agency. Almost all interest groups viewed Ruckelshaus as an acceptable choice for EPA leadership. American business receptivity to him was considerable, despite the prediction that he would renew interest in pursuing violations. In part, the response reflected a conclusion that Ruckelshaus would bring greater stability and certainty to regulatory programs. In part, industry and environmental groups were impressed with administrative experience, whatever the implications for them. Ruckelshaus's appointment telegraphed the message that other factors that promote compliance could be mobilized. His resignation in late 1984 created consternation among environmentalists and unease among businesses which welcomed the professionalism he had brought to environmental policy.

In addition to general leadership qualities (Kagan, 1984), other attributes of senior government officials influence their ability to promote compliance. Administrators with long tenure in an agency can establish a credible reputation among the regulated. Seniority also has other effects; an agency veteran is usually better able to ascertain the correct balance of sanctions and incentives to apply to different classes of violators. And a competent, experienced manager can amass the resources—ranging from the latest monitoring equipment to qualified lawyers—necessary to mount an effective program. Commitment to

[3] *Inside EPA*, Apr. 9, 1982, at 6.
[4] Interview A-1.

the agency by its chief managers tells industry that it will encounter the same government personnel over time. Challenges to the regulator may be unwise.

Some studies (Bernstein, 1955; Sabatier, 1977) identify a cycle of decay of effectiveness of regulatory agencies caused in part by administrators' developing overly sympathetic attitudes toward industry.[5] But capture sometimes results because administrators conclude that they must look to the private sector for career options. If government service is a career, this influence may not apply.

Agency Characteristics

Whether for reasons of absence of coordination, incompetent leadership, or excessive workload, some environmental agencies are constantly in turmoil. Priorities for implementing laws are not set; knowledge of enforcement capacity is limited or nonexistent; rule development is constantly off schedule; and the nature of appointments to critical midlevel positions reflects no serious commitment to environmental objectives.

Agency limitations may derive from the legislative tendency to transfer numerous demanding tasks to the bureauracy without supplying sufficient implementation resources. EPA administers or has administered the Clean Air Act, the Energy Policy and Conservation Act, the Safe Drinking Water Act, the Noise Control Act, the Federal Insecticide, Fungicide, and Rodenticide Act, parts of the Atomic Energy Act, the Uranium Mill Tailings Radiation Control Act, the Resource Conservation and Recovery Act, the Toxic Substances Control Act, the Clean Water Act, the Marine Protection, Research and Sanctuaries Act, the National Historic Preservation Act, and Executive Order 12114. Implementation of each of these laws requires dozens or even hundreds of rules, reviews, and decisions. One TSCA provision, relating to premanufacture notification for new chemicals, requires EPA to review four notifications per day while at the same time seeking to process analyses of toxicity of the 60,000 chemicals already on file.[6] Similarly, to choose but one example under the Clean Air Act, in 1981 EPA had fifty-eight regulations under consideration. These ranged from revising the priority list of New Source Performance Standards to altering emission standards for commercial aircraft to reduce hydrocarbons, carbon monoxide, and nitrogen oxide.[7]

Some difficulties derive from the agency's inability to mount a convincing technical effort (Roberts & Bluhm, 1981). The *Distler* case[8] highlights this

---

[5] See discussion of capture theory in Chap. 6.

[6] Mannix Letter to the Editor, *Regulation* 3 (March–April, 1984).

[7] *Federal Register,* Jan. 14, 1981, Part IX, Environmental Protection Agency, Agenda of Regulation.

[8] See Chap. 1.

problem in the enforcement sphere. The state was slow in developing a program for control of toxic substances. Kentucky is the site of the infamous "Valley of the Drums," ranked high on the EPA's list of the most dangerous hazardous waste disposal sites. The state had actually purchased equipment for the scientific tests required to link pollution residues to polluting sources, the legal task in *Distler*. But it never used the equipment. Criticism of the efforts of some air quality management districts also centers on technical weaknesses. Several are common: devices used to monitor compliance are not in working order; inspectors fail to employ equipment made available to them to detect pollution properly, instead relying on "their eyes and noses"; and inspection schedules are unprofessionally (or illegally) communicated to pollution sources indicating when compliance will be monitored.[9]

Industry also assesses inspector professionalism, and performance of agency personnel in the field influences agency reputation. The total regulatory process at times appears to be nothing other than the person in the plant at the time. The inspector has been characterized as a "surrogate for the rule itself,"[10] and Clay (1983) and Bardach and Kagan (1982) report that his or her professionalism is a correlate of the efficacy of enforcement.

Fundamentally, agency incompetence and nonprofessionalism raise the question of how to comply. Many regulations are neither clearly stated nor self-executing; government must supply supplemental information about acceptable response. Often these messages about the nature of standards and about the ways of manifesting and reporting compliance lack precision. Deficiencies in agency performance can also seem to make the very act of compliance optional; if the agency is incapable of minding its own affairs, chances of its systematically pursuing noncompliance are small. In light of poor government performance, the business community can move from a defensive to an offensive posture and insist that agency management makes compliance unrealizable.

During the tenure at EPA of Anne Gorsuch, *The Wall Street Journal* reported that industry became increasingly critical of EPA administration. One automotive lobbyist noted: "More and more businessmen say they are concerned about the lack of direction in leadership."[11] Nonprofessional behavior of several

---

[9] *Los Angeles Weekly,* Feb. 29–Mar. 5, 1982, at 9, 58.

[10] Comments of Steven McDonald, at California Air Resources Board, Air Pollution Enforcement Symposium May 24–26, 1983.

[11] *The Wall Street Journal,* Apr. 7, 1982. These comments sound mild, however, when juxtaposed with some criticisms by environmentalists. Anthony Roisman, formerly chief of the Department of Justice section of hazardous waste in the Land and Natural Resource Division, speaking before a house subcommittee on oversight and investigation of the Energy and Commerce Committee, concluded:

> Anne Gorsuch deserves our pity rather than our contempt as she struggles to carry out the orders delivered from the White House," Roisman said. "The vaunted 'Ice Queen' is more an 'Ice

kinds was reported. Industry concluded that there was no set environmental policy, and companies could not identify agency personnel with whom to negotiate. A vice-president of the Motor Vehicle Manufacturing Association described another result: "The agency staff is afraid to make decisions because they aren't getting clear signals from the political people at the top."[12] An indirect result feared by industry is that in the absence of leadership in the agency Congress might articulate more stringent environmental requirements than the administrator would promote or might reassert standard-setting authority that it had delegated.[13]

Nonprofessionalism has other correlates. If the agency does not convincingly present intelligent background for its policies, enforcement activities and other attempts to effect compliance will encounter continuous challenges. For example, as is elaborated in Chapter 6, in the early 1970s it was common knowledge in the corporate sector that the EPA based some of its conclusions about the health effects of environmental pollutants on shoddy science and insufficient data. An automobile executive summarized that standards were based on "a perfectly miserable data base."[14] Furthermore, "scarcely one bit" of data had been added in the last half decade to the health studies that were done for mobile source emission standards. "The health evidences are weak, the atmospheric data is negligible, and the benefits are nonexistent."[15] In these circumstances the probability of a successful challenge to a regulation is high, and the regulatee is quick to generalize from one vulnerable area to other control programs.

A requirement that is difficult to achieve if not impossible has limited influence. Overly strict standards and unrealistic deadlines become the rationale for general attacks on environmental policy. "Infeasible" becomes "impossible" in the industry assessment.

Industry no doubt overuses arguments of impossibility, but even a few impossible-to-achieve goals can detrimentally affect government's reputation. A classic case in the early 1970s involved water quality standards. Officials of the Michigan Water Resources Commission were chagrined when they learned of a judicially set requirement to apply a standard that was physically impossible to meet: "The Court changed the effluent requirements: It did not have a technical

---

Puppet' in these political maneuvers." [He added that [Gorsuch] "has been forced to barricade herself in the EPA penthouse to avoid dealing with an agency in open rebellion, losing two of her top aides to "personal reasons" in six months, still unable to fill key assistant administration positions on her staff, [and] unable to control the flow of confidential agency data to the press. . . . Ms. Gorsuch is totally out of her depth–a perfect Charlie McCarthy for the Reagan/Bush Edgar Bergen. *Inside E.P.A.*, Apr. 9, 1982, at 6.

[12] *The Wall Street Journal*, Apr. 7, 1982, at 29.

[13] Interviews for this study, conducted early in the first administration of President Reagan, cited the inaccessibility of Ms. Gorsuch.

[14] *Environment Reporter*, at 1838 (1977).

[15] *Environment Reporter*, at 1838 (1977).

understanding: one parameter was lowered: one highered. . . . [It] . . . imposed an oxygen requirement above what the water can hold.''[16] A more global example is the continued application of the 1987 standards under the Clean Air Act to regions which will not reach these air quality goals in the 1980s. Such cases suggest other situations wherein the impossibility defense can be used. There is a further impact: when government concedes that standards cannot be met, the definition of acceptable compliance is problematic.

Agencies can develop highly professional reputations which promote compliance. Access to resources for development of rules and for adequate enforcement may help, but large budgets and staffs are not a *sine qua non*. Agency reputation is a function of the use of available resources and the husbanding of symbols of legitimacy.

Administration that appears to be informed, fair, efficient, and smooth generates respect for the regulator. A review of both more successful and less well regarded administrative actions suggests several types of performance that impress both the regulated and the general observer. The integration of scientific information into rule making enhances legitimacy when the agency uses respected experts (Kagan, 1984) and requires or sponsors quality research, and yet recognizes the need to take action even when knowledge gaps exist. Stakeholders in a public agency also appreciate the careful consideration of a variety of points of view in regulatory matters. Agencies are more highly regarded if their procedures are understandable and accessible and not subject to endless continuances. Clear and unequivocal articulation of rules and enforcement decisions is impressive. And regulatory conclusions reached upon a balancing of several interests enhance legitimacy. In situations which are scientifically unclear the agency should appear to make decisions on close calls in ways that promote the public interest. The agency's reputation will benefit if it successfully communicates the fact that decisions represent the least costly way of achieving important societal objectives.

Observers will doubtless differ on assessments of the legitimacy of agencies and on the overall effectiveness of administrations, but all agree that general classifications can be made. In recent years Ruckelshaus's decision on lead was well respected, whereas the EPA failure to list formaldehyde under Section 4 of TSCA and the CHESS conclusions in the 1970s were actions that weakened the regulator's reputation.

The Administrator as Ideologue

From both the conservative and liberal ends of the political spectrum come attacks on agency professionalism focusing on the political philosophies of ad-

---

[16] Interview D-7 in DiMento (1976).

ministrators. Critics conclude that government officials fail to meet their responsibilities as managers; instead, they promote ideological goals. Put with equal force by detractors of Democratic and Republican EPA administrations, the EPA was described as directed either by those overly sympathetic to industry or those who placed environmental goals far ahead of other important societal objectives. Criticism of Gorsuch was vitriolic during the last months of her administration; it centered on her allegedly rewarding political conservatives at the expense of environmental programs. Management problems allowed little time for regulatory activity.

Many industries concluded that agency performance during the Carter administration was ideology-driven. An industry spokesman described business resistance to regulation and antipathy to government's presentations of the pollution problem: "If there were noncompliance the assumption was made by government that industry wanted to dirty the air, kill people in autos and use a lot of gas."[17] An official holds "a press conference so that every major newspaper in the country hears him say that the auto industry is going to kill babies and old women. We simply don't think that's fair."[18]

Other businessmen describe government administrators who would "restructure society":

> They . . . have really a quite different philosophical outlook from the classical conservationist organizations. They're out to remake the country, the society, they're much more radical in their approach. And they're much more coercive in their means. They're adept in using the law in the political system as it exists.
>
> There's a group . . . who really see this as an opportunity to recast our society in a better mold. They are not Marxists . . . my guess is that if you could ever pin them down on a philosophy it would be one . . . very akin to fascism except for the bad features. . . .
>
> I think they are utterly sincere. . . . I think . . . they feel they're just as good citizens as I would feel I am a good citizen. . . . [They are] basically young people who are less concerned with freedom because they . . . have been regimented most of their lives anyway. . . . I think their picture is one of a benevolent autocracy.[19]

A comment by Senator Edmund Muskie of Maine in a congressional debate on the Clean Air Act, quoted by the court in *Chrysler v. EPA* (see Chapter 1), indicates early perceptions of industry by some environmentalists:

> We cannot solve the problem of whatever technology the industry chooses to put its bets on. All we can do is set the standards. The automobile industry has created all of the problems from the top to the bottom.[20]

Thomas A. Murphy, then chairman of General Motors, noted (Gatti, 1981):

---

[17] Interview D.
[18] Interview D.
[19] Interview B.
[20] 116 Cong. Rec. 33096, Leg. Hist. 335.

> This country urgently needs more statesmen in government and fewer adversaries, fewer anti-business zealots. I am thinking particularly of those who seem bent upon always casting business in the role of villain, as the source of all our country's troubles . . . such single-minded people . . . are capable of great damage. This country can ill afford the bitter price of the adversary climate these zealots have committed themselves to preserve and to advance. We in American industry cannot afford to tackle the world marketplace without our government on our side. We simply cannot compete effectively against foreign manufacturers who are being cheered onward by their governments who view them as prime national assets, if at the same time our own government treats its domestic industry as an enemy, to be hampered, held back, whittled down. (p. 142)

Perceptions of an ideology-driven administration counter compliance when they highlight and exaggerate general government–business differences over regulation. Attention to the value of specific rules—which may themselves be relatively noncontroversial—is replaced with ongoing debates about political theory and government's proper function. Government exacerbates the rift by failing to recognize that some decisions may be truly trivial and that flexible, pragmatic responses can win over potential adversaries.

The Quality of Lawyering in Environmental Litigation

> If you could imagine a lawyer . . . as passive, insecure and lacking both any knowledge of environmental regulations or chemistry and no ability to examine or cross-examine, you'd have (the defendant's) lawyer, Mr. X. Put Mr. X against Z, young, articulate, well-versed in environmental regs and chemistry and toxic waste, and a Board of well-educated, impatient men . . . and you could easily predict the outcome of the hearing as well as the humiliating way Mr. X was treated.[21]

Lawyering is another important aspect of agency professionalism. Lawyers to a large degree determine and control the translation of words of new environmental statutes and rules into action forcing incentives and commands. Law is carried out by the legal profession; to make pollution control law efficacious, effective lawyers are essential. Within government agencies and in regulated businesses (as discussed later in this chapter), specialization in environmental law, experience in negotiation and litigation, and organization and administration of the legal unit all contribute to the perception of government professionalism in achieving compliance.

Environmental law practice grew rapidly in the 1970s; but not every agency has access to this expertise, and the quality of available assistance in this field is as variable as in any. A large percentage (51%) of surveyed enforcers reported that resources were insufficient for their enforcement activities, and each of these officials noted that the agency lacked personnel for pursuing cases. Nonetheless,

---

[21] Observations by monitor of proceedings in *Bob's Plating* (see Appendix for complete citations).

the quality of legal representation was not reported to be a major problem. Forty-five percent of enforcers considered legal representation to be as good for government as for the private sector, whereas 21% considered it inferior. Thirty-eight percent of those surveyed thought that the legal representation available to government was excellent; one-quarter of those studied (24%) had concerns about it.[22]

The influence of good and bad legal work manifests itself in many ways. Prosecuting attorneys may not be able to articulate clearly to the court the rationale for a sanction for an environmental insult or the legal theory that makes that insult a violation. Even in cases in which the facts are favorable to the environmental plaintiff and well-developed legal concepts are applicable, inexperienced lawyers can incur adverse judgments.

Legal defenses for environmental violations often manifest considerable resourcefulness. However, when government chooses to use its most experienced litigators and focus resources to prosecute a serious violation, results will usually be positive. General compliance will also be promoted as enforcing agencies acquire strong litigation reputations.

Distler, Frezzo, and *Chrysler*[23] were all well litigated by government, and in *Derewal*[24] lawyers for the United States Department of Justice eloquently argued the admissibility of critical evidence that the defense tried to exclude on the basis of privacy. In a colorful analysis of judicial precedent from the automobile search exemption of the fourth amendment,[25] the government stated:

> In the instant case, there is virtually little or no reasonable expectation of privacy in the contents of a chemical tanker-trailer, a commercial vehicle, which is in the process of being dumped into a public body of water while located at a public place. The fact that the contents of such vehicle while privately owned are those with a very limited "privacy" is shown by the requirement that such vehicles contain documents describing the nature of their contents. Furthermore, the fact that the defendants were engaged in discharging the contents of the tanker into the river is a direct disavowal of any expectation of privacy in its contents.[26]

In *Derewal,* the defendant was sentenced to six months imprisonment, fined $20,000, and placed on probation for four and one-half years for illegal discharge of pollutants into the navigable waters.

In *Frezzo,* an appellate court upheld a jail term on a negligence theory for the first time in environmental law. The government pursued the case vigorously

---

[22] See surveyed enforcers, discussed in Chap. 2, p. 36.

[23] See Chap. 1.

[24] *Derewal* (see Appendix for complete citation).

[25] *Chambers v. Maroney,* 399 U.S.42 (1970); *United States ex rel. Johnson v. Johnson,* 340 F. Supp 1368 (E.D. Pa 1972): and *United States v. Chadwick,* 97 S.Ct. 2476 (1977).

[26] *Derewal,* government's Memorandum of Law III, May 9, 1978.

and established three principles which have far-reaching influence in pollution control litigation: (1) a showing of willfulness could be made on circumstantial evidence; (2) criminal liability can be imputed in a pollution control case despite the absence of evidence that the defendant was present when the allegedly criminal action occurred; and (3) a severe sanction can be imposed on a theory of negligence.[27] Law review commentary (Note, 1980) concluded that the case was good precedent for greater criminal enforcement of the Clean Water Act.

Many other examples of competent and creative legal work are found within public agencies and within those public interest law firms which act as environmental watchdogs and catalysts. During the environmental decade the Department of Justice was as respected and feared in environmental, public lands, and natural resources cases as it was in many other divisions. Its attorneys, which with agency counsel pursue EPA cases, were among the most competent in the environmental bar. But it remains the case, perhaps to a somewhat lesser extent than in the 1950s and 1960s, that both government agencies and environmental interest groups are places where many lawyers do their post–law-school learning. Federal and state environmental agencies, NRDC, and EDF are training grounds as well as career opportunities. Turnover remains relatively high, and the insights and techniques developed in cases promoting compliance may be effectively employed later in cases defending industry, as the environmental lawyer "graduates" to a major corporation or to a large defendant-oriented law firm. Moreover, the states, especially the smaller ones, do not have the luxury of selectivity enjoyed by the Department of Justice.

Lawyering is crucial both when regulators apply traditional legal concepts to new fact settings and when they employ new environmental statutes. In each situation, government lawyers face judges who are likely to be either deeply steeped in traditional concepts of property or who are genuinely ignorant of the philosophical thrust of the new environmental law. An assertive educational process is demanded of government lawyers, sometimes in the form of arguments that the uninitiated are hesitant to make.

When government first used counts under the Michigan Environmental Protection Act (MEPA)[28] in legal complaints, uncertainty about the reach of the statute was considerable. MEPA language does not clearly direct organizational action, creating a cause of action to protect "air, water and other natural resources and the public trust therein" from "pollution, impairment or destruction." The contribution of MEPA to Michigan's enforcement capacity thus depended largely on the aggressiveness with which both state and private attorneys developed the law's meaning, sometimes in highly unusual factual situations (Haynes, 1976; Sax & DiMento, 1974). The act was read to cover the

---

[27] Interview with Bruce Chasan, Jan. 8, 1985.
[28] Michigan Comp. Laws Ann. §§ 691.1201–.1207 (Supp. 1973).

issuance of permits to drill exploratory oil and gas wells in a state forest;[29] to
protect sand dunes against mining abuses when no other statute was violated;[30]
and to prohibit burial of animals contaminated by toxic substances when no
public health provision was involved.[31] MEPA also was litigated to enlarge the
reach of other Michigan laws.[32] Similarly, the meanings of several of the phrases
in the Clean Air Act and its regulations became potent only because government
and public interest lawyers effectively asserted a far reaching interpretation.

Bad lawyering can become part of the mythology that forms the reputation
of an agency, and regulated industry can conclude that probabilities of successful
prosecution are small. The following internal memorandum became public during the later months of the Gorsuch administration at EPA:

> It has come to my attention that there is a growing amount of lateness and informality
> in attire by the enforcement division attorneys. We are all professionals, and should
> hold ourselves accordingly. For this reason, I would request that all enforcement
> attorneys come to work properly attired; this would include jackets and ties for all
> male attorneys. In addition, the practice of arriving at work late and taking two hours
> or more for lunch must stop immediately. We are being paid for a full day's work and
> this is what is expected of us.[33]

Quite apart from the questionable logic of the relationship between dress and
legal excellence, the memorandum and other reports reflected a demoralized and
frustrated department whose effectiveness in securing compliance was problematic. Weak personnel can undermine agency influence. Industry regularly
assesses an agency's lawyers, technical experts, and policymakers and makes
implicit and explicit determinations about probabilities of meaningful
enforcement.

Support Groups: Advocating Compliance

Enforcers and regulated industries do not operate in a political and social
vacuum on questions of compliance. In some areas of criminal law, professional
prosecuters perform their duties even when citizens are indifferent to violations;

[29] *West Michigan Environmental Action Council v. Natural Resources Commission*, 405 Mich. 741,
275 N.W. 201 538 (1979).
[30] *Lincoln Township v. Manley Brothers*, No. 001113-CE (Cir. Ct. Byrns) Dec. 20, 1984.
[31] *Board of Commissioners of Kalkaska County v. State*, No. 74–619-CE (Cir. Ct. Kalkaska County,
July 4, 1974).
[32] See *Michigan State Highway Commission v. Vander Kloot*, 392 Mich. 159, 220 N.W. 2d 416
(1974) (involving the state Highway Condemnation Act); *Superior Public Rights, Inc. v. DNR*,
No. 73-15862-CE (Cir. Ct. Reisig), Mar. 2, 1976 (involving the State Department of Natural
Resources' procedural duties); and *Little Wolf Lake Property Owners Association v. Haase*, No.
74-000837-CE (Cir. Ct. Glennie, July 8, 1975), applying MEPA to the Subdivision Control Act.
[33] *Inside EPA*, Mar. 12, 1982, at 4.

such is not the case for much of regulation. Collective concern generally is a necessary condition for government action. Called by many different names among the disciplines (pressure, influence, political involvement), support for adherence to rules channels the activities of agencies to promote compliance.

Support groups generate and transfer information to both regulators and regulated businesses. Their function in our compliance framework may be to create new information, to repeat or highlight information already available to government and business, or to fill in gaps in the regulatory process which reflect information deficiencies. They set the backdrop for business–government interactions. They are the entities which care about either special or general compliance.

Support groups important to compliance are those with a major interest in environmental quality or its concomitants. Relevant groups include those whose primary purpose or organizational goal is economic development—as well as collectivities promoting air and water quality and the like. We address in this section both compliance promoting and compliance obstructing or mitigating third-party activities.

Support groups may be large incorporated entities with sizable national membership, such as the Pacific Legal Foundation or the Sierra Club; of these, sufficient members exist to fill a directory. Or small *ad hoc* groups that coalesce around a local environmental controversy may promote support for regulatory action. What is essential to influence is not size but the capacity to communicate consistently and clearly views on compliance and noncompliance to decision makers in government and in the firm. Our use of the term *support group* does not match precisely the term *interest group* in political science, for our use covers groups that are both smaller and larger than interest groups and those that might have shorter life spans. In all areas of regulation support groups are of critical importance, and in the short history of environmental law they have been especially influential.

These support groups communicate the severity of violations, priorities for enforcement, and the relative importance of other societal objectives, including some which may counter environmental compliance. Means of influence vary with the groups and with phase in the regulatory process. Influence may come at the legislative and rule-making stages through lobbying for one form or another of rule or through campaign contributions. Support groups may work through the courts and target important environmental cases to force favorable interpretation or implementation of regulations. Groups build support for compliance through litigation programs the purpose of which is to create precedent for forcing institutional compliance. Through various means—among them selective citizen suits and formation of coalitions with others—groups enhance their individual messages. When committed to compliance objectives they can quite dramatically alter the vigor with which rules are interpreted and the vigor with which law is enforced.

Outside groups which support environmental objectives and occasionally company insiders can play central roles in implementation of law (Ball & Friedman, 1965; Dolbeare & Hammond, 1971; Milner, 1971; Muir, 1967; Roberts & Bluhm, 1981; Sabatier, 1975). Groups which have strong attitudes about compliance influence law's behavior out of proportion to their numbers (Roberts & Bluhm, 1981; Schwartz & Orleans, 1967; Wichelman 1976), and potential victims of environmental violations are uniquely positioned to promote agency attention to violations (Kagan, 1984, citing Sabatier & Mazmanian, 1984). Surprisingly, treatments of business response to law often ignore this "pressure for compliance."

In addition to the previously cited modes of political participation, support comes in the form of simple visits to regulatory agencies and regulated firms directly and through education campaigns. Groups can move a prosecutor to initiate an action against a noncomplying business. Continuing concern for the outcome of litigation can influence the zeal with which government pursues a case, the period for which it pursues an action, and the manner in which the regulator treats an alleged violator.

Support groups help establish priorities for the numerous violations observed and observable by government. Support group activity indicates the importance of a regulation to its intended beneficiaries (Roberts & Bluhm, 1981). Statements by attentive groups may be more responsible than the general fiats of legislative bodies. The case selection process is a means of achieving efficiency when government is overloaded.

Identifiable community and national groups can help form a firm's view of the law (DiMento, 1976, Sabatier & Mazmanian, 1978). Support can make salient to top company officials a matter that might otherwise be treated as marginal, avoidable, and trivial. Clinard and Yeager (1980) suggest that the extent to which enforcement is felt, understood, and appreciated by high-level corporate executives is important to its deterrent effects.

"Elite" groups, local *ad hoc* groups, and the persistent and determined individual can direct the outcome of suits and the implementation of environmental policies (DiMento, 1977; Dolbeare & Hammond, 1977; Geis, 1978; Muir, 1967; Wilson, 1980; Wirt, 1970). EPA studies (1980) conclude that support groups prompt government to enforce a directive. Several of the most dramatic pollution control victories of the 1970s resulted from legal actions of environmental groups.[34] Both procedural and substantive law have resulted, with effects ranging from enjoining environmentally destructive projects to developing a new judicial orientation toward review of administrative agency action.

---

[34] *Citizens to Preserve Overton Park, Inc. v. Volpe*, 401 U.S. 402 (1971); *Scenic Hudson Preservation Conference v. Federal Power Commission*, 354 F.2d 608 (1965); *Calvert Cliffs' Coordinating Committee v. Atomic Energy Commission*, 449 F.2d 1109 (1971); *Natural Resources Defense Council, Inc. v. Morton*, 458 F.2d 827 (D.C. Cir. 1972); and *TVA v. Hill*, 437 U.S. 153 (1978).

The function of support groups in new regulatory arenas is particularly important because for the most part compliance norms are neither fully articulated nor widely shared. Agency officials look to the most vocal government stakeholders in determining how and whether to implement and enforce law. And as Roberts and Bluhm (1981, p. 382) note, regulators "assume, perhaps correctly, that for every active citizen they see, there are numerous others who share the same viewpoint."

Such agency behavior is offensive to some government observers: using the "squeaky wheel" to monitor implementation of law assertedly violates principles of representative democracy. Classical political theory offers no good rationale for basing regulatory enforcement on interest group complaints. The point is made here descriptively; the facts remain that the modern liberal state is so actively involved in citizens' affairs and guiding laws are so numerous that government administrators must seek means of setting priorities. The political and legal environment of pollution control law now resembles the criminal justice system: law enforcement is not a function totally of a legal philosophy nor is it based on a pure theory of government. Enforcement is much more pragmatic, responsive to the ever growing and often changing demands of the citizenry.

Added to this practical explanation of the importance of support groups in fostering compliance are factors addressed elsewhere in this book. First, statutory language is not sufficiently precise to make law self-enacting or compelling. Second, legislators typically do not know exactly what they mean; they legislate to avoid specificity that will alienate some constituencies. Third, representative government churns out laws at a pace that its administrative arm cannot fully process.

Enforcement efforts of the California SCAQMD illustrate support group influence. For example, a lively group of business neighbors got the agency hearing board to refrain from issuing (as it was inclined to do) another compliance extension to a small boat-painting company. The neighbors cited continuous problems of odor which affected their own commercial attractiveness; reportedly, the board would have bowed to the painting company's pleas had it not been for the attendance of these interested parties.[35]

Support group power is enhanced in some areas of public law by evolving moral standards. Corporate and business practices once seen as part of normal business competition now qualify as criminal—if not heinous—as the effects of violations become appreciated. Yet assessments are not always shared by a large part of the population, and therefore communication, by a few, of norms, moral outrage, and concern with enforcement can define how government will proceed (Gunningham, 1974).

Finally, support groups assist compliance in another way. In circumstances

[35] Interview J.

in which a target desires to comply, for example out of economic self-interest, the support groups can provide a rationale for compliance to its customers, constituents, or stakeholders who oppose compliance. Business can "blame" compliance on actions outside of its control; these include suits brought by the support group or threats that the support group makes. An example in the civil rights area involved the equality in public accommodations provision of the 1964 Civil Rights Act. Southern businessmen saw the integration of their restaurants and lodging facilities as a source of increased revenue. Business pointed to the federal government and civil rights groups as creating the need to integrate. Thus they could open their doors to blacks while appeasing white customers by maintaining agreement with racist sentiments (Champagne & Nagel, 1983).

Coalitions for Compliance

Combinations of groups which support compliance can be particularly powerful. Organized environmental protection or consumer groups may link up with local groups. The resources and skills of established organizations complement the passion and unchoreographed public worry, fear, and outrage of individual victims of an environmental mishap. In *Narmco*,[36] which was resolved by a *nolo contendere* plea by the company, a group of mothers was concerned about the effects of the company's emissions on the health of children and on property values. California's Campaign for Economic Democracy gave technical assistance to the group. The impetus for the mothers' organizing had been a chemical explosion in August, 1979, which resulted in the death of one worker and serious injury to another. The group submitted two dozen citizen complaints describing dust, soot, odors, fumes, coughing, allergies, bronchitis, and other respiratory conditions in adults and children—graphic testimony which is not likely to go unnoticed by a trier of fact.

Expert assistance, such as by university researchers, can increase the influence of citizen initiatives. Epidemiological work often effectively contextualizes complaints of a neighborhood group; specific health problems described by locals are put into a framework that decision makers can evaluate. Local officials also enhance the status of support groups by incorporating environmental goals in their political platforms. A case in point was a California community effort to have air quality regulations enforced against an oil refinery which admitted ongoing noncompliance. Citizens advocated enforcement in well-attended meetings in Santa Fe Springs, where the refinery (Powerine) is located. Present at public hearings were stalwarts in grassroots political activities, academics from the School of Public Health of the University of California, and a populist councilman who declared that "no facility should be able to expand without the

---

[36] See Chap. 6.

proper permits. . . . Now I don't care if it's Mobil, Powerine, or Joe's Gas Station.''[37] Numerous local environmental organizations now recognize the value of working with at least one elected official in pursuit of violators.

Citizen suits brought by environmental action organizations may have considerable leverage in compliance efforts. The long list of important regulatory actions taken subsequent to litigation or pressure by the EDF, NRDC, Friends of the Earth, and the Sierra Club includes clarification of the meanings of compliance with provisions of the National Environmental Policy Act and the formulation of several rules under the Clean Air Act, the Federal Water Pollution Control Act Amendments, the Toxic Substances Control Act, and the Resource Conservation and Recovery Act. Much of the activity of the environmentalists has focused on holding agencies to legislative timetables for rule promulgation and program establishment.

The relationship between support group influence and compliance is not always positive. Support for individual enforcement activities can complicate procedures for promoting compliance. Citizens may set priorities in ways that fail to recognize overall agency objectives, especially in situations in which resources are limited. They may disrupt opportunities for cooperative resolution of environmental problems. Or they may divert agency attention toward violations or alleged violations with limited significance. Most disruptively, support groups can inject incorrect information into the compliance activity; they may inaccurately describe a firm's actions and lead government to investigations that uncover no true derelictions. An enforcement officer described the latter problem:

> Information I received [on alleged violations] was normally fragmentary. People complaining about a problem would state the facts in a way to make the problem seem as bad as possible, leaving out many facts that could put the situation in another light. Further, most of our information came from members of the public who did not understand the law. Often they would not have the relevant facts.[38]

Citizen suits can also pull agency attention and resources away from long-range planning and act as obstacles to building an overall compliance promoting capacity (DiMento, 1976).

One administrator in a state water resources agency criticized citizen environmental suits:

> In technical matters such as water pollution control, proposals by the court should be broad and the specifics determined by administrative experts. This would be a better way to get a desired end result. If there are deficiencies in administrative law, these should be concurrently changed and reformed because courts are not able to do these

---

[37] Councilman Escondrias, Aug. 19, 1982, at Santa Fe Springs Advisory Council Public Hearing.
[38] Surveyed enforcers, discussed in Chap. 2, p. 36.

things in an expendient manner. If we had to rely on courts, we'd be far behind; it's a
cumbersome process to rely on courts.[39]

Bureaucrats in another agency expressed concern over citizen disruption of agency programs:

Now there are people and I'd even call them "Johnny come latelies," who are
concerned about the environment who want to grab hold of this program and do a lot
of . . . things with it. They don't know how fundamental, how basic, how productive
the program has been. Compared with what there used to be there's tremendous
progress.[40]

Business Support

Community support can also affect outcomes of environmental controversies to favor noncompliance. Put differently, support for economic objectives
can mitigate environmental enforcement. Local response to an environmental
incident is a barometer of importance that regulators might attach to the event. Of
course, the relationship between being liked in the community and treatment in
the criminal justice system is not linear. Were it so, few prosecutions of neighborhood gamblers and other violators of the traditional "sin crimes" would be
successful. But environmental law enforcement based on national ideological
support of compliance is often mitigated by local understandings of an offense
and familiarity with the defendants. Gathering support for prosecution of some
classes of violations is difficult, especially when would-be defendants are residents of tightly knit communities and where negligence, at worst, caused the
environmental problem.

The industrial sector can do much to counter environmental enforcement
and build support for its activities. Influence is greatest when in the form of an
undivided front of a large industry, as opposed to even persistent voices of
individual companies. An EPA spokesman summarized in the case of the proposed phase-down of the use of lead in gasoline: "It is not sufficient for a couple
of companies to protest. . . . EPA is looking for united industry action."[41]

Business opposes—in legislative hearings, in the popular press, and in
community meetings—policies allowing environmental legal action. Industry
typically asserts that liberalized rules would create a business environment forcing companies to move or close down. Such predictions are repeatedly made in
hearings on bills that would give citizens a greater function in environmental law
enforcement:

If this becomes law any individual may file a lawsuit against any corporation of any
kind in the State of Michigan whether public or private for a real or fancied wrong

---

[39] Respondent D-23 in DiMento (1976).
[40] Respondent A-2 in DiMento (1976).
[41] *Inside EPA,* May 25, 1984, at 2.

resulting from actual or fancied pollution. If every individual is permitted to file for damages, every do-gooder, kook, psychopath, or malcontent will file so many actions that our already overburdened judicial system . . . will be in danger of breaking down.[42]

A company may tie a threat to an individual lawsuit. For example, the Ford Motor Company stated that it might need to close its doors forever if the Pinto case were successfully prosecuted. The case saw Ford charged with reckless homicide and criminal recklessness for the fuel system design of the Pinto. Ford was indicted by a county grand jury in Indiana subsequent to the burning death of three teenage girls who were involved in a low-speed, rear-end collision in their Pinto. Ford was acquitted in a jury trial.

American Business: Correlates of Compliance

Compliance studies have also paid attention to the nature of the firm. Work has centered on several questions: Rather than enforcement techniques, rather than external social conditions, are there characteristics of firms that best explain who complies and who does not? Are there "good apples and bad apples" (Kagan, 1984)? Are there determinants of compliance structurally intertwined in private companies?

Research focuses on both the individual firm and the target industry. Size of the firm (Child, 1973; Clinard & Yeager, 1980; Diver, 1980; Lane, 1953; Mitnick, 1981), its economic profile (Clinard & Yeager, 1980; Henderson & Pearson, 1978) and the general economic health in the industry,[43] the organizational structure of the company (Roberts & Bluhm, 1981), and the business culture have all been linked with performance and number of violations.

Size

Organizational size is relevant to compliance. Scholars use size as an analytic factor in various ways including number of employees, average assets, gross revenues, number of divisions, and volume of production. Although firm size may help explain noncompliance in highly circumscribed situations, law and society theories do not agree as to whether largeness relates to greater or lesser compliance. The subjects studied differ as do units and indicators for comparison; and some explanations center on the actual commission of violations

---

[42] From a tape of the State of Michigan Senate debates on the Michigan Environmental Protection Act, June 26, 1970.

[43] Some business respondents felt that their positive profit profiles made them tempting targets for enforcement action, but this factor was ranked "not at all important" by 45% of the enforcers surveyed and "very important by" none. Surveyed enforcers discussed in Chap. 2, p. 36.

whereas others describe government's motivation to uncover or pursue existing noncompliance.

Diver (1980, p. 291) suggests that "penalties are proportionately less severe the larger the size of the enterprise or the more serious the risk of harm presented." Penalties do not create a great incentive to comply. Furthermore, large businesses are better able to mold the legal environment in which they operate. And in certain complex industries, regulators find it difficult to identify the nature and causes of the noncompliance problem and to locate the "levers of control" which might be used to minimize noncompliance. Finally, the larger the firm the greater the difficulty in determining for whom criminal *mens rea* will be judged and, more generally, whom to name as defendants.

On the other hand, prosecution of larger organizations may be of greater public interest than pursuit of small companies, and therefore the large company may be more legally exposed (Clinard & Yeager, 1980). Prosecutors may find them more attractive targets. Government may base choices on analysis of the most effective use of enforcement resources, or its response may be political. Political interests derive from aggregate citizen concerns or may represent strategic government decisions to promote an individual or agency objective. Victories over large companies are more valuable politically than successful litigation involving small businesses.

Others explain why small size may correlate with environmental violations. Demands on regulatory agencies, including quotas or requirements that a specified number of enforcement actions be initiated, may lead government to concentrate on small firms. The regulator may consider chances of success, measured by convictions and consent orders, better (Kagan, 1974, Stigler, 1970) and the small business adversary as weaker or less politically threatening. Poor violation records of small firms relative to large companies are also explained in several ways. Hawkins (1980) concludes that large organizations can afford to comply; other studies (Mitnick, 1981) also cite their access to costly expertise, the need for large capital investments, and the economic benefits of compliance.[44] The latter include both savings linked to process changes and profits created by domination of markets so that smaller firms cannot compete successfully if they comply. A related finding is that poor financial performance, as measured by profitability, efficiency, and liquidity, is related to overall risky behavior and to illegal activity (Clinard & Yeager, 1980); however, environmental noncompliance does not fit the overall patterns. Clinard and Yeager found that pollution control violations are more prevalent in firms within industries which enjoy favorable profit trends. Nonetheless, "violating corporations on average tend to have lower growth rates than the nonviolating companies" (p. 130); and this result did apply to environmental violations.

---

[44] See "3M Gains by Averting Pollution," *Business Week*, Nov. 22, 1976, at 72.

Large firms with long-standing reputations in a community also may wish to maintain a profile as "good neighbors" and respond to directives that newer companies with fundamental survival goals and those preoccupied with developing profitability may choose to ignore or confront. Representatives of the "smaller large automobile industries" have made this point, observing that the smaller auto companies may lack sufficient resources to develop compliance systems found in the largest American corporations.[45] Similarly, the smaller waste disposer often poses a greater threat than the well-funded and highly visible larger company.

The SCAQMD profile of the companies most difficult to control reflects a combination of small size and concomitant lack of sophistication about the regulatory process. Among the most unresponsive, even to informal and personalized attempts to promote compliance, are dry cleaners, service stations, and water heater manufacturers. Basic communication difficulties compound the compliance challenge for some firms. Many of the employees of these firms do not speak English[46] and clearly do not understand the letter or spirit of some environmental rules.

Differentiation

Size correlates with differentiation in the firm (Mileti, Gillespie, & Eitzen, 1979). Differentiation refers to functional complexity: the degree to which tasks are subdivided for performance by individual groups within a business. Companies can be differentiated by production process, by inward-looking activities such as maintenance and support, and by those that scan the external environment, including units which aim to influence regulation and respond to legal challenges. No clear linear relationship between compliance and differentiation exists. Nonetheless, the literature on differentiation offers several important insights on compliance.

One group of studies links high differentiation to both firm noncompliance and to government's inability to effectively promote compliance. Gross (1978) points out that companies faced with "turbulent" environments differentiate and those departments most exposed to societal demands are likely to "exhibit deviant behavior." Vaughan (1982, p. 1393) has noted that "size and the complexity that frequently accompanies size provide many locations in which unlawful behavior might take place." Units within firms that have little interaction with the outside world may be able to hide noncompliant activities (Mitnick, 1981).

---

[45] Interview D. See also Mitnick (1981) specifically citing the Chrysler Corporation as facing this problem.
[46] Interview K.

Locating employees responsible for assuring compliance (Schelling, 1974) is difficult in highly differentiated businesses.

Even when the expressed goal of the firm is to implement compliance protocols, coordination of responsible units may be difficult (Ermann & Lundman, 1975). The highly differentiated firm may also lack clear authority lines for bringing about compliance and meeting regulatory standards. For example, the Conference Board found that a high percentage of corporate decision makers in environmental affairs has ''no relationship'' to line operations (Lund, 1977).[47] Thus high differentiation may produce violations as standard outcomes of individual procedures, each of which is legally blameless. Or high differentiation may see one unit promote noncompliance without knowledge of top management (Roberts & Bluhm, 1981).

On the other hand, differentiation may promote regulatory compliance if firm structure includes influential units with a self-interest in positive response to pollution control regulation (Mitnick, 1981). Support within the firm for environmental law may result. Mitnick also notes that differentiation may dampen the link between profit seeking and rule avoidance:

> Where ownership and management are separated, fragmentation and high dispersion of ownership exists . . . managers may not be policed to perfectly attain the owners' profit goals. The self-interest goals of managers may then be served in part by adherence to regulation and by responsiveness to constituencies that favor regulation. (p. 73)

Differentiation and Lawyering

Differentiation in American business usually means development of specialized legal divisions. Again, no single view dominates analysis of the relationship between the characteristics of the legal function of a company and compliance. Legal divisions may see regulatory compliance as their *raison d' être* (Mitnick, 1981) and favor responsiveness to law. On the other hand, firms with legal expertise may hear only bad news from counsel about the nature of regulations, and at times lawyers exaggerate the potential for government intervention. The legal department can thus become isolated (Gross, 1978). To many managers in a firm, the lawyers represent an obstacle to business progress. Businesses may screen the bad news of legal requirements from those at operating levels and even from corporate executives. That bad news includes the need to meet health and safety and environmental standards.

But good lawyers can skillfully define compliance in ways acceptable to the firm and postpone enforcement. Highly differentiated firms can capitalize on

---

[47] The study also found that a sizable majority (88%) of its respondents lodged environmental policy decisions at the level of company vice-president or higher.

environmental law specialization. Fundamentally, conclusions about compliance are legal determinations. In interactions with government, lawyers can mold the working interpretations of control law (Lund, 1977). They can lessen regulatory disruption by channeling and narrowing law's impact. Simpler firms will not have this husbanding function. Businesses without specialists must rely on outside counsel in critical cases, putting the firm in a very vulnerable position.[48] Companies without environmental law capacity (and these include some of the largest American corporations)[49] may fail to employ a legal defense which is obvious to a specialist. Inexperienced counsel may not recognize technical definitions of compliance whereas experienced lawyers can undermine even well-developed enforcement actions.

A significant example is the application of bankruptcy law in defenses of environmental wrongdoing. In *Chem Dyne*,[50] the individual defendant had been successfully enjoined from hazardous waste dumping and ordered to clean up existing wastes. Following a failure of the company to comply, the state of Ohio appointed a receiver to collect from the individual defendant to finance cleanup. The defendant then filed a personal bankruptcy suit. Ohio in turn sought a hearing on his finances, whereupon the defendant moved the bankruptcy court for an injunction against Ohio proceeding in state court. Ruling against the state, the bankruptcy court concluded:

> There is no difference in substance between efforts to collect money from a debtor by securing a court order, and efforts to enforce a money judgment against him. We hold, therefore, that the state is bestopped to deny that it is seeking a money judgment against a debtor.[51]

As the government has argued, use of the Bankruptcy Reform Act[52] can operate to block the effective implementation of a mandatory injunction obtained in environmental actions.

Informal communications initiated and carried on by company lawyers with the government increase the cost of promoting compliance. Business counsel can complicate an initially straightforward case (Diver, 1979). Requests for clarifica-

---

[48] Both a large company with a relatively nonspecialized legal division and a small business reported this vulnerability. Executive interviews discussed in Chap. 2, p. 36.

[49] Reported one automobile manufacturer in an interview for this study: "Our senior attorney is not versed in environmental law" (Interview D). Strategic advice of the most experienced lawyers is not always followed. Advice to oppose (during the rule-making process) the selective enforcement audit rule under the Clean Air Act—a rule which later became a major source of problems for the auto industry—was not followed by one manufacturer (Interview A).

[50] *In re Kovacs*, 29 B.R. (S.D. Ohio 1982).

[51] *Id.*, see also 15 ELR 20121 (U.S. Jan. 9, 1985) wherein the United States Supreme Court affirmed that the "site operator's obligation under an injunction to clean up its site can be subject to discharge in bankruptcy."

[52] 11 U.S.C. § 362.

tion, for conferences to share new information and for extensions, and descriptions of good faith efforts to realize compliance act as obstacles to the timely realization of government's objective. Demands by industry lawyers may lead an agency to put aside a prosecution to pursue easier tasks and those that can readily be accounted as agency victories.

Law as "Stone Wall"

The firm with extensive litigation resources can also use legal maneuvers to slow down enforcement. Jeffrey Miller, a former head of the EPA enforcement division, has described some of these tactics.[53] Initiating broad litigation on matters related to the enforcement action "could be used either to divert government resources from its original goal of suing the company or to make it clear that trifling with the company can be costly to the government." Information requests can slow down the government effort: "Use the Freedom of Information Act to gather as much information as possible about the government's case, its evidence, its related policies and procedures, and the scientific and technical justification for its conclusions and demands." Aggressive discovery obstructs enforcement: government "employees are not well-disciplined and may well reveal information damaging to the government's case. . . . Its records are poorly kept and it will have difficulty in complying with exhaustive requests for documents . . . giving the judge a poor impression of the government's case and professionalism."[54] The well-represented firm can also employ preemptive litigation, resisting requests for information and quashing warrants to discourage or overwhelm an environmental agency.

Standard Legal Procedures and the Challenge of Compliance

These examples of strategic use of law to promote client interest and descriptions of government and business legal capacity and procedures introduce a general issue in the analysis of compliance. The large number of lawyers and their central places in the lawmaking and enforcing institutions of the regulatory process may be sources of tunnel vision on achieving compliance; this phenomenon may counter the overall realization of environmental goals.

The dominance of lawyers in Congress and the state legislatures, in the administrative agencies, in the public interest firms, and in influential offices in American business contributes to an unbalanced situation. Command and control regulation and the counterpart of industry legal response is the standard operating mode of addressing environmental problems, despite the availability of other

[53] *Inside EPA,* Apr. 2, 1982, at 13, quoting Jeffrey Miller.
[54] *Inside EPA,* Apr. 12, 1982, at 13, quoting Jeffrey Miller.

approaches for developing rules and ensuring that they are followed. Choice of strategy is seldom based on thoughtful analysis of the inventory of compliance-promoting strategies presented in Chapter 3.

Several modes of dynamics in lawyer-dominated institutions impede general compliance. The self-interest of some proportion of the bar cannot be ignored; lawyers do very well under traditional regulatory forms. But other explanations also exist. Much of the education of lawyers focuses on adversarial methods and characterizes problems as win–lose controversies. Until recently, few law schools have included offerings on alternative dispute resolution procedures in the curriculum and even fewer have required such courses. Indeed, the content of regulatory courses has emphasized the impotence of conference and conciliation, citing problems that arise from too cozy relationships between government and business. Furthermore, recruiting for law schools and into legal careers necessarily selects those who thrive on competition and the strong advocate.

The criticism that legal institutions and lawyers obstruct compliance with environmental objectives is an interesting hypothesis. But at this time little of the critique of the legalization of gaining compliance is based on convincing empirical results or makes its case by means of comparative analysis with other compliance promoting regimes. Anecdote and case study dominate work addressing both the costs of excessive legalization and the benefits of consensus-seeking schools. Nonetheless, dominance of the regulatory process by lawyers who have been trained to litigate, who emphasize the benefits of adversarial systems, who know primarily command and control (except in certain areas of labor law and the like) suggests that part of the noncompliance problem results from blindness to processes of negotiation and mutual problem solving—methods more prevalent in countries such as the United Kingdom.

A Word on Corporate Culture

What has been loosely called corporate culture influences compliance (Clinard & Yeager, 1980). Corporate culture refers to a company's reward structure, its propensity for risk taking, its attitudes toward growth, its emphasis on product or process, and the extent of its assignment of responsibility for company actions to formal groups.[55] Whether an enterprise approaches issues and problems through use of a bureaucratic and formal structure or personalizes issues is relevant to compliance. Some organizations work through committee mechanisms for several activities, including the monitoring of regulatory activity and the formulation of the company's response to proposed law. For instance, General Motors has been known for its use of committees and its considered and

---

[55] "Business Strategies for Industries in Transition to Deregulation," course announcement, Wharton School Executive Education Program, May 15–19, 1983, Wharton School, University of Pennsylvania.

managerially conservative corporate reaction to new ventures and to criticisms of government.[56] The business community has described recent shifts in GM's orientation, directed by Chairman Roger Smith, as surprising and extraordinary.[57] Such shifts may lead to adventurousness that creates additional enforcement targets. A greater willingness to attack regulation aggressively and defend vigorously against enforcement actions may also result. Chrysler presents itself to the world both in company publications and unofficially[58] as more personally oriented, its executives as more casual about formalities and likely to decide policy individually. It is a corporation whose actions have been called "aggressive and visible."[59] Chrysler appears to revel in a role as corporate David taking on government Golaith in the regulatory sphere, and its executives conclude that because of this posture the company may be more vulnerable to enforcement. Ford likes to be seen as the obedient, socially responsible citizen; it will make every effort to educate government on the impacts of regulation but sees itself as complying fully—no matter what the costs—once proposed rules become law.

Cultural effect may derive from factors that were important at an organization's founding, as Roberts and Bluhm (1981) found in their study of the utilities:

> TVA continues to see itself as the low-cost-power yardstick that Franklin Roosevelt envisioned. Ontario Hydro still puts a high priority on Adam Beck's goal of reducing Ontario's dependence on the United States for energy. (p. 373)

Or it may derive more recently from the strong ideologies of critical top management people. In this regard, Clinard's findings (1983) on middle management are relevant; those company officers promoted from within, particularly in family-dominated enterprises, are said to be more willing to conform to law. In contrast, managers hired in were found to be less concerned with ethics and with strict legal compliance and more concerned with development of their individual reputations and with professional mobility. This difference may derive from the personalization of noncompliance that may result when violations occur in companies "with the family name on the door."

Corporate culture mediates reaction to policy initiatives and provides a guide for managers to respond to environmental regulations. One sees an Exxon developing a journal on environmental analysis and ecological issues, an ARCO bringing together environmentalists and industry and government representatives

---

[56] J. B. Schnapp, *Corporate Strategies of the Automotive Manufacturers* (Lexington, KY: Lexington Books, 1979).

[57] *The Wall Street Journal,* July 2, 1984, at p. 17.

[58] Interview A-2.

[59] *The Wall Street Journal,* Jan. 1, 1983, at 3.

for seminars to analyze business–government relations, and a Mobile or Gould waging a belligerent journalistic campaign against the ideology of government controls of the market. Over time, these reactions become part of a company image, shared by executives and employees, and drive reaction to change.

CHAPTER 8

MAKING ENVIRONMENTAL LAW WORK

Introduction

Compliance has no one definition. Noncompliance has no one explanation. A farmer or small businessperson sees noncompliance as a response to unfair regulations promulgated by bureaucrats who do not possess adequate knowledge of the private sector. Legal counsel to an environmental agency concludes that the pollution problem would be solved if government more efficiently and aggressively enforced major regulatory laws. The vice-president for environmental affairs of a large corporation, a resident of a community that has been polluted by a neighboring industry, the technical consultant to a legal suit involving unacceptable disposal techniques all have different views of the compliance problem.

Recommendations aimed at achieving compliance must recognize the varying and at times contradictory perceptions of rule violations and must take into account the complex processes which make for noncompliance. Models based on simplified understandings of the behavior of business firms, such as those which dominate the law and economics literature, do have a heuristic value. Grand theories about the impossibility of effective regulatory regimes also help us to be realistic about the degree of obedience to law that one can expect. But pragmatically policy makers are better served by suggestions which recognize, but are not overwhelmed by, the variety of individual, organizational, and systems dynamics that determine compliance.

The recommendations that follow recognize the perspectives of the various participants in the compliance event. Yet the suggested policies also reflect awareness of incompatibilities among some objectives, and therefore priorities

163

are set among the goals of public policy. The suggestions are aimed at achieving more valuable, not necessarily more compliance. "More valuable" here means compliance that makes a measurable difference in environmental quality and takes into account costs and other social objectives.

We mean the recommendations to be attainable in the short run and to be pragmatic. The recommendations stop short of radical legal and institutional reforms, which we consider unnecessary.[1] They are based on the premise that noncostly and nonintrusive ways of achieving acceptable compliance exist.

The findings of this study and the thinking of American businesspeople and of those who practice and comment on environmental law underpin the recommendations. Suggestions aim to improve the effectiveness of groups which support the goals of meaningful environmental law. A common premise of this chapter is that public policy should direct attention and resources to those violations that society has collectively—although assuredly not always consensually—determined to be a priority problem.

Other common themes pass through the recommendations. First, each suggestion recognizes the power of good information, given other necessary conditions, to drive decisions and actions. The suggestions combine to promote availability of quality scientific and everyday information. They underscore the need in the regulatory process to create, translate, and communicate usable, instrumental information. They aim to reject information which would mislead or misdirect the regulatory process and to protect the regulator from bad or irrelevant information. The recommendations assist those who promote the flow of usable information among support groups, government, and industry. Second, the recommendations recognize that information alone, no matter of what quality, may be impotent in the face of resistance to compliance. The recommendations thus use select institutions to create pressure and incentives for compliance. Third, the reforms attempt to maximize motivation to comply: many of the suggestions envision a kind of multiplier effect of individual interventions.

Making Environmental Law Communicate

Operational Law

Regulations cannot stop at fiat. To squeeze a statement of wonderful environmental objectives through a legislature in a furious session will not by itself change the behavior of a polluting firm. Central to reforms in environmental law is operationalizing the regulatory program. By *operationalizing* we mean thinking through the stages of implementation of law, specifying behaviors to be

---

[1] Stone (1975, at 228) states that his recommendations involve "more widespread invasion of corporate managerial autonomy than anything ever tried in this country."

touched by the rules, analyzing costs to those who will be benefited and those who will be burdened, and establishing timetables for performing tasks.

An expression of a societal goal in an environmental statute is only a beginning of successful environmental control; the minutely detailed behavior required of regulatees is the end. In between are many acts and decisions to be taken by identifiable but not always easily identified groups and organizations. What process or activity is to be controlled? When? And, sometimes, how? These questions should be raised at the agency level if the legislature has delegated lawmaking authority. If important steps in the regulatory path are missed, what contingencies will apply? How is the desired behavior best induced or coerced? How will follow-up be undertaken if monitoring demonstrates that compliance is not forthcoming? Upon discovery of noncompliance, what sanctions or incentives will be used?

Regulatory responsibilities must be fully described; admittedly this is a large task. Who ultimately will be affected and in what ways is not always obvious in the development of regulations. But environmental policy should attempt to identify probable sites for business response and the activities that need to be promoted, enhanced, or curtailed to reach environmental goals. Development of scenarios may help assure comprehensiveness of rules. Regulators can use such scenarios, as in the case of an operation like Frezzo's, to test the sensitivity of regulations to constraints on compliance. What is the likelihood that this class of agricultural producers would understand this type of water quality rule? What types of changeover would such rules require at what cost and with what benefits?

Some regulations do recognize this need to consider operational detail. Federal rules on treatment and disposal of low-level radioactive wastes, for example, identify responsible parties for overseeing wastes for hundreds of years and specify actions required of them.[2] In each stage, rules identify those will have the lead role in control and note their obligations. Later generations of NEPA guidelines also reflect this attention to meaningful communication. Reformers also based NEPA changes on other principles of communicative law addressed later in this chapter.

Thinking through tasks required to implement environmental law is similar to activity required in any form of effective management. Without operational detail, directives can move into a vacuum. High-level government and business managers upon first assuming office often are surprised when their orders translate into little more than expressed desires. Requiring that control equipment be affixed to pollution streams, that manifests be created and filed, that plant conditions ensure against environmental mishaps are tasks much different from promising environmental quality improvements or ordering the executive lunch.

---

[2] 10 CFR 61. Another important example is the Hazardous and Solid Waste Amendments of 1984, P.L. No. 98-616, 96 Stat. 3221.

Total information about the nature of regulated settings never is available. We will never be fully aware of the variety of production processes, control mechanisms, and interactions among chemical and other substances that make up the immense industrial base. We will never completely understand the idiosyncracies of organizations which mediate their capacity to comply. Therefore complete articulation of rule implementation will be difficult. However, when undertaken in concert with other recommendations for efficacious regulatory law, rule making which manifests an understanding of regulated settings and employs means of upgrading information about the firm will promote general compliance. Resulting regulations will exhibit an intelligence about regulated settings and will avoid the vagueness in law that makes compliance difficult and erodes respect for regulators.

Full Communication and Its Relationship to Other Reforms

A focus on operational law provides an opportunity to consider adoption of a variety of environmental reforms. Two offered during the last decade, integrating regulations across media and customizing rules, may serve as examples.

EPA has considered for several years the multimedia approach to achieving environmental quality. The ultimate aim of a multimedia program is to combine controls on pollution of air, water, and land to effect the greatest environmental results with available resources. Rather than moving the disposal problem from the air to the water and then on to the land—and perhaps into the groundwater—the strategy may provide ultimate sinks for pollutants.

Multimedia programs require analysis of costs and benefits associated with disposal in each of the media. These analyses have not been completed partly because of other single-medium demands on regulatory agencies and partly because knowledge about ecological effects of disposal routes has developed slowly. Until a full systems analysis is undertaken no multimedia rule can be promulgated. But that analysis is unlikely unless environmental regulation is reformed to provide a respite from other regulatory demands, and incentives to do multimedia work will be effective only if the reforms are given a realistic life span, as described below.

Customizing or particularizing rules can also be given a fairer hearing if government works to operationalize law. Regulations can reflect knowledge of the regulated industry—location, production processes, access to less polluting sources, availability of substitute materials, local ecological characteristics, and pathways for movement of pollution. And rules can capitalize on decentralized sources of information about environmental problems. Presently, under the Clean Air Act,[3] industrial development is differentially allowed on the basis of

---

[3] 42 U.S.C. § 7401–7642.

regional air quality classifications. Nonattainment areas are treated differently from those which meet NAAQS, and pristine areas are distinguished from those which have experienced development. But more localized or micro-environmental customizing is possible, going beyond such distinctions as new and old sources in the major federal acts; criteria could be developed which classify firms according to their potential contribution to pollution within a region and which determine whether a plant will be able to generate its own compliance standards.

Customizing rules is sometimes associated with business design of the regulatory framework. If properly supervised, self-auditing and self-regulation might promote compliance. The EPA is gathering information on existing audit programs and providing a clearinghouse for information sharing on this strategy. The agency is also analyzing the idea with emphasis on review of existing programs and experiences, their costs and benefits to business and government, and the experience of the self-auditing counterparts in the Securities and Exchange Commission, the Food and Drug Administration, and in European countries.[4] EPA proposed a sulphur dioxide ambient standard that reflects a receptivity to customizing. The standard would take note of the fact that the sulphur content of coal varies, as do meteorological conditions of regions where coal is burned; controls would be demanded in terms of these differences.[5]

To assure that customizing, such as at the plant, is not abused, an environmental control agency would have to gather considerable information about the firm, have a strong backup enforcement program, and evaluate success by means of overall performance guidelines. As Braithwaite has acknowledged (1982, p. 1492):

> There are many areas where the dangers of cosy local agreements would be intolerable. However cooptation can be controlled in many cases by a particularism severely constrained by overreaching standards which were themselves products of a national debate.

Recommending operationalizing of rules is not the same as promoting either multimedia controls or customizing. But the increased opportunity to program environmental regulations is also a chance to consider several kinds of environmental law reforms, including ideas which have not been adopted precisely because it is not clear how they would be implemented.

To improve regulations, lawmakers should more intelligently employ cost–benefit and cost effectiveness analyses—tools already used in the regulatory process.[6] EPA has taken initiatives in this area, for example, for NSPS under the Clean Air Act.[7] The analytical aids should be applied with attention to an

---

[4] *Inside EPA*, Apr. 8, 1983, at p. 6. See also 50 Fed. Reg. 46504 (Nov. 8, 1985).

[5] *Inside EPA*, Sept. 17, 1982, at p. 5.

[6] See the Federal Trade Commission Improvements Act of 1979, 15 U.S.C. 58 and Executive Order 12291, Feb. 17, 1981.

[7] *Inside EPA*, Jan. 6, 1984, at p. 4.

understanding of the groups which are affected by rules and with sensitivity to the limitations of cost–benefit analysis (McAllister, 1980; Swartzman, Liroff & Croke, 1982). Integrating these tools into rule making would leave less of the prediction of regulatory outcomes to assumption (Lave, 1981). Costs should include the full range of primary and secondary expenses in the short and long run. Benefits, too, should be comprehensively defined. Economic analysis must include activities often underemphasized in control activities, especially operations, such as worker training and maintenance of control equipment. Cost–benefit work which goes beyond the preconceptions of agencies about a proposed action is necessary.

Because of the many limitations on cost–benefit analysis, its capacity to objectify and precisely to quantify environmental decisions is limited. Interested parties to environmental regulation will not agree on discount rates to employ, factors to consider and to monetize, and other technical matters (McAllister, 1980; Swartzman *et al.,* 1982). We recommend applications of this analytical tool which emphasize the specification of factors that are employed in decision making and which underscore trade-offs among regulatory choices. Seldom will the cost–benefit analyses transcend value choices in policy making or translate to a numerical outcome that is useful; such a number is not our aim. Disciplining regulators' choices and making those choices more visible to the regulated is our objective.

Attempts to operationalize rules have a concomitant outcome to substantive results: they force the regulator to communicate more fully with the regulated firm. Otherwise, the breadth and depth of information for good studies will be lacking.

Setting Priorities

Support groups can imagine an almost infinite number of regulatory programs. Each group considers its goals noble. To improve regulations and increase the probability of overall compliance with results, citizens must answer some basic questions: Am I willing to trade achievement of some objectives for realistic and effective controls elsewhere? Which regulations are most important? Which do we address now? Which can be achieved by realizing other objectives?

Priority setting is highly value-laden. However, approaches exist which recognize this social aspect of the regulatory process and address the dilemma of participation in a context in which regulatory centralization, expertise, and finality are also needed. Scoping is helpful prior to official attempts at lawmaking; it is a procedure first utilized in environmental law in the late 1970's. In scoping, groups which have some stake in a proposed regulatory program are brought together to discuss and evaluate it. For example, in a review of proposed changes in the Clean Air Act, government officials, environmentalists and industry representatives assembled to assess the wisdom and impacts of provisions on ambient air standards, prevention of significant deterioration of air quality, nonattainment area strategies, and technology forcing.[8]

Stakeholders have also addressed the effects of acid rain in scoping sessions. In 1982, industry and government officials developed statements of, among other things, research needs on acidification of the nation's waters. If consensus about research can be reached, receptivity to regulations based on results of scientific study may follow. Among the participants were the federal Interagency Task Force on Acid Precipitation, the American Petroleum Institute, the Electric Power Research Institute, General Motors, Ford, the Coordinating Research Council, the United States Department of Agriculture, the National Oceanic and Atmospheric Administration, the EPA, and a dozen states.[9]

Scoping identified both areas of agreement and disagreement over regulatory priorities. No guarantees exist that consensus will evolve, but differences can be used in priority setting. One veteran of a scoping meeting noted: "My personal expectation is that it is very unlikely that these sectors would join hands . . . [but] . . . ways to meet the needs of the other groups" were recognized.

As the examples in Chapter 3 demonstrate, environmental negotiation and mediation also can provide detail for operational law. Both scoping and negotiation processes recognize the need to identify parties that have a stake in the outcome of a decision. Both are concerned with proper representation of affected groups and with the need to address value issues explicitly in complex environmental choices. Both underscore the need for greater information sharing to create a context of acceptable outcomes; both admit that different policy routes can arrive at similar ends; and both evolve out of a frustration with the weaknesses of command and control and litigation pathways to promoting compliance.

Other procedural innovations recognize appropriate expertise lodged in government and in the private sector and allow for sharing of perspectives on regulatory priorities. The Health Effects Institute (HEI) holds some promise for developing scientific analysis. The EPA and the automotive industry jointly established the institute, which is governed by an independent board with impeccable credentials.[10] Its mission is to bring together leading scientists to define and carry out research on the health effects of mobile source emissions. The institute establishes research priorities and solicits proposals from outside groups. HEI then evaluates results and reaches conclusions regarding regulatory implications. The initial research agenda for the institute included, among other

---

[8] *Inside EPA,* Nov. 26, 1982, at p. 4.

[9] *Inside EPA,* July 24, 1981, at p. 6.

[10] The Board of Directors in 1985 consisted of Archibald Cox, Carl M. Loeb, a professor in the Harvard Law School; William Baker, Chairman Emeritus, Bell Laboratories; Donald Kennedy, President, Stanford University, and formerly administrator of the United States Food and Drug Administration; and Charles Powers, President, Clean Sites, Incorporated.

projects, study of the cardiovascular and other health effects of carbon monoxide and susceptibility to respiratory infections related to exposure to nitrogen oxides. When information outcomes are as instrumental as those HEI contemplates, opportunities for meaningful business–government information sharing and cooperation are significant.

The consensus workshop can also generate regulatory specific information. The workshop aim may be a quantitative risk assessment or a set of conclusions about the health effects of a substance. Participants are industry, union, government, university, and research institute scientists who seek consensus on regulatory issues such as whether formaldehyde should be regulated. Workshop members have ready access to research findings. The formaldehyde workshop[11] addressed the health effects of the substance through panels on epidemiology, exposure, behavioral and psychological effects, sensitization to the compound, metabolism, carcinogenesis, and risk assessment.

The workshop idea is embryonic, and several weaknesses exist; these include considerable probability of gaps in the data base, possible bias in selection of experts, absence of standardization of the rules which apply to evaluating the quality of scientific work, disagreement on the appropriate application of scientific judgment, and procedural defects which threaten full explication of available information. But with time and experience the consensus workshop and similar institutional innovations (Brooks, 1984) may provide a useful forum for sharing government and industry perspectives on the scientific bases of environmental rules and creating information applicable to regulatory needs.

We also recognize the limitations of consensual, cooperative approaches to making operational law. For certain environmental values, compromise is unacceptable to one group or the other (Amy, 1984), such as OSHA's attempt to negotiate a benzene exposure level;[12] and the approaches will be futile. But much of the environmental agenda is not so controversial. Tactics which involve actors in the compliance framework hold considerable promise for improving the communicability of, respect for, and ultimately the enforceability of law.

Operational Law and Elected Officials: The End of Liberalism Revisited

Recommendations which call for increased cooperation among government and business and other interest groups, which advocate negotiating, mutual priority setting, and consensus seeking must be clear about the essential role of law in a regulatory framework. Participatory mechanisms which aim to enhance

---

[11] *Consensus Workshop on Formaldehyde,* Little Rock Arkansas, Oct. 2–6, 1983.
[12] "Environmental Mediation Hits Snag," *Chemical and Engineering News* 15 (Oct. 15, 1984) at 15.

the quality of information in regulation must not be confused with bargaining that obstructs the legitimate and necessary use of the police power.

One of the most articulate analysts of the evils of uncontrolled participation is political scientist Theodore Lowi. Lowi (1969), in an indictment of interest group liberalism, describes the results of government policy which substitutes bargaining for the rule of law, mistakes abdication of standard setting for democratization, and ratification of special interest agreements for consultation with collectivities which will be affected by law.

Lowi warns that the failure of government to fulfill its legitimate lawmaking and standard-setting functions leads to policies which are mere sentiments, destroys political responsibility, and makes politicians out of bureaucrats, those government officials who should be holders of usable expertise. Lowi calls for several reforms to reinstate the rule of law. The holding in the *Schechter* case should be revived: the courts must again declare invalid and unconstitutional delegations of power to administrative agencies when those delegations are not accompanied by clear standards. Second, the agency must require rule making early and apply it frequently. Third, government should create a senior civil service, insuring that an independent generalist class of dedicated administrative officials guides policy making. Fourth, a tenure of statutes act is necessary, making all organic acts reviewable and subject to repeal every five to ten years.

In summary of his argument, Lowi says:

> A good clear statute puts the government on one side as opposed to other sides, it redistributes advantages and disadvantages, it slants and redefines the terms of bargaining. It can even eliminate bargaining, as this term is currently defined. Laws set priorities. Laws deliberately set some goals and values above others. (p. 126)

Lowi's warning is important to evaluations of greater participation by the governed, of responsive law, reflective regulation, self-auditing, and other cooperative strategies. The costs of unchecked involvement can be real. Yet the indictment is compatible with the reforms called for here.

One must recognize the possible inverse relationship between bargaining and perceived legitimacy of government and its rules. But information sharing, legitimately channeled and visible to all interested in regulatory outcomes, does not mean bargaining. Called for in the operationalization of law are government efforts—sometimes in the legislative branch, sometimes administrative—to make rules as clear as possible and to set standards (not to create bargaining positions) based on comprehensive intelligence of the regulated sector. Sought through participatory means is not necessarily the final rule but information that makes for a fully informed ultimate choice. Sought through participation by the enterprise is not policy making, is not shared exercise of the police power, but fair and full articulation of perceived costs and benefits of proposed rules. Sought is information not normally available when government operates in an exclusively adversarial or command and control form. In making law operational through interactions with the firm, government must not confuse healthy flexibility with informality that equates business's contribution with governing. In a period of infatuation with public–private agreements and of frustration with government's ability to set clear guides and standards, Lowi's presentation of the evils of uncontrolled interactions is timely.

Operational Law and Officialdom

Should legislators themselves be involved in operationalizing law? Locating responsibility for increased clarity and specificity of law in the legislature is desirable in certain areas of environmental regulation. On some policies, value differences are so serious and perceived environmental and economic effects so great that elected officials should be directly involved in rule making. Decisions on nuclear waste disposal[13] and on controls of substances about the health effects of which scientific analyses diverge are examples. For the latter, legislative action is especially useful when economic impacts of controls would be disruptive. The legislature's active participation can enhance law's status as well as its communicability. The benefits of communicability may be considerable if those who conceive of broad environmental objectives are also required to address means of reaching them.

Nonetheless, because of the volume of environmental and health and safety problems, requiring legislators to specify the details of law may have an overly conservative effect. Demanding active involvement of elected officials could obstruct regulation in some areas. Lawmaking that moves outside of legislators' areas of technical competence may also be misguided (Ackerman & Hassler, 1981). Therefore legislative standard setting should be limited to the most difficult cases. For other environmental programs renewed and fuller use of legislative oversight could suffice; and greater assertion of authority by competent administrative officials can promote the rule of law and the development of clear and efficacious standards. The nature of administrative involvement is addressed in further detail later in this chapter.

Operational Law: Some Conclusions

Programming rule implementation, cost–benefit analyses, negotiation, and priorities in regulations—reforms aimed at improving communication of law— generate fairly significant front-end regulatory costs. Rule making will take longer and involve more stakeholders. But potential benefits are also considerable. Agencies will be able to direct resources and time to formulate fewer

---

[13] DiMento, Joseph, Lambert, William, Suarez-Villa, Luis, and Tripodes, James. "Siting of Low Level Radioactive Waste Facilities," *Journal of Environmental Systems* 15, 1 (1985–86), at 19–43.

environmental control strategies, and they will be able to draw on additional expert assistance for rule formulation. Targets should perceive resulting regulations as more professional and legitimate. Because industry and small business participate more meaningfully in the creation of rules, they will treat results as more authoritative than those generated without active involvement.

Limiting the scope of regulation allows the regulatee a more considered response to law. Although the conclusion is difficult to substantiate, the regulatory environment of the 1970s may have been so turbulent that full compliance can be seen as an Herculean success rather than an ordinary response of competent American businesses. We need not concede that compliance was physically impossible, as some industry analysis would suggest, but for some firms regulatory chaos followed the rule-making schedules that federal, state, and local governments announced. Organizational studies recognize the need for special managerial skills during such periods (Weiner & Mahoney, 1981). These skills are not common, and it is not good public policy to expect the exceptional throughout industry.

Analysis of communication of information in the regulatory process points to three more immediate and tangible recommendations. They address the information flow to administrators, the structure and staffing of agencies, and the information exchange between business and government.

Information Flow to Agencies

More immediate reforms relating to communication of regulation are necessary. Except in situations wherein public policy explicitly embraces more informal procedures, such as in negotiated rule making, *ex parte* communications in standard setting must be controlled (Preston, 1980; Verkuil, 1980). *Ex parte* communications are those that occur outside of the record that an agency builds. They may come in the form of written materials submitted after government closes a comment period or in the form of personal contacts not subject to analysis by other parties.

Rules against *ex parte* communications aim to insure that an interest group "has a last clear shot" at information that is aimed at influencing those in authority. The formaldehyde and dioxin cases[14] demonstrate how *ex parte* communications can undermine formal protections in rule making which aim to deliver the best possible information base to the decision maker. In these cases, conclusions supposedly based on scientific analysis were injected into regulatory policy making ignoring the basic premises of scientific inquiry and of opportunity to respond.

Control of *ex parte* communications becomes more challenging when infor-

[14] See Chap. 6.

mation outside the record is offered across agencies or from the executive branch to the agency. Some observers argue that the executive always has the option of providing new information to its agencies—after all, the agency is but an extension of the White House or of the governor's office. But although it is true that senior officials at EPA and in other regulatory agencies work at the pleasure of the president, it is also true that the Administrative Procedures Act[15] guides the agency in rule making. The better approach appears to be that which subjects executive missives to the same rules that guide communications from other interested parties. The public should be able to scrutinize and comment on information that is to be background for standard setting and for policy decisions about whether to regulate a substance.

If government adopts negotiation and other administrative reforms the purpose of which is to improve the information base of rule making, the concern with *ex parte* communications is somewhat different. Creation of a formal record may inhibit the frank, detailed, and exploratory exchange among participants in mediation and negotiation. And, ultimately, informal communications may enhance the information available before the final rule is promulgated. Yet concerns remain over the rights of those who are not party to consensus-seeking procedures. No simple resolution exists, but, as noted in Chapter 3, commentators have devised some possible procedural safeguards. Further experiments will determine whether the safeguards are effective or whether they so complicate the administrative process as to negate the benefits of the original informal thrust.

Independent EPAs?

Making environmental bureaucracies independent regulatory agencies is worth further consideration. Theoretical and political analyses of performance differences of independent regulatory commissions and regulatory agencies within an executive branch are extensive (Mitnick, 1980; Rabin, 1979). One presumption has been that independent regulatory commissions are simply that: independent and less responsive to short-term political concerns.

Yet, in considering reforms one must acknowledge strong criticisms of the commission form. Mitnick (1980) warns of isolation of the commission "from active, corrective public constituencies" and resultant capture by the regulated industry (p. 69). He asserts that independent commissions may also be removed from important social, economic, and administrative changes "to which other agencies respond" (p. 69). In the last category are management techniques that

---

[15] 5 U.S.C. §§ 551–559; 701–706; 3105; 3344; 5362; 7521 (1967). For a rather complete analysis of the practices required in agency reaction to new information on the record see *Ethyl Corporation v. EPA*, 541 F.2d 1(1976).

other forms of government may pick up more easily. Isolation is related to absence of accountability, a problem which various advisory groups, including the Ash Council in the Nixon administration,[16] have recognized. The Ash Council also concluded that there is less legislative oversight of commissions than there is of executive agencies; and that commissions have more difficulty in managing and in recruiting top professionals than do administrative agencies with single heads. Finally, others have observed that the commission form would inhibit coordination with other agencies whose missions are affected by EPA activity.[17]

Empirical work on any of these propositions has not been persuasive (Moe, 1982; Rabin, 1979; Robinson, 1971; Stewart, Anderson, & Taylor, 1982). Nonetheless, *prima facie,* changeover to a form of the independent commission improves regulations on several dimensions. Contemplated results include greater continuity in policy, greater consistency in regulatory programs, more time to test regulatory innovations fully, and less chance that short-range political needs will dominate environmental policy. Under an independent commission,[18] the chief executive (president or governor) would appoint members who, if confirmed by the legislature, would serve staggered terms, longer than the four-year term of the executive. Yet the executive office would continue to provide an avenue to accountability through its control over the commissions' budgets and appointment of commission chairs.

A related recommendation is the fuller employment of a senior civil service in agencies, much like the pattern found in Great Britain. Professionalism and continuity of policy at the highest levels of environmental agencies are the goals. Political appointees moving in and out of important agency positions—even at midlevels—have little incentive to assume a long-term perspective. In the United States shifts in policy-making echelons produce regulatory results that are confusing to business.

In certain periods the weaknesses of political appointments have been egregious. Two cases give a sense of the extent of EPA problems in the first half of the Reagan administration. One EPA official was terminated for lying to a Congressional inquiry[19] commission about her attempts to fire an employee for whistle blowing. Others were linked with attempts to set priorities in hazardous

---

[16] The Advisory Council on Executive Organization headed by Roy L. Ash, "President's Advisory Council on Executive Organization," *Presidential Doc.* (Feb. 8, 1971), at 174, cited in Glen O. Robinson, "On Reorganizing the Independent Regulatory Agencies," *Virginia Law Review,* 57 (1971), at 947–995.

[17] Reed, P., "The Environmental Protection Act of 1983—Is an Environmental Protection Commission Necessary?" *Environmental Law Reporter, 13* (Mar., 1983) at 10064–10065.

[18] The reform has already been introduced in Congress. See H.R. 2362, 98th Cong., 1st Sess. (Scheuer), Mar. 23, 1983.

[19] *Los Angeles Times,* Feb. 5, 1983.

waste clean-up activities on the basis of political prospects of candidates in the polluted region.[20] If a truly professional Securities and Exchange Commission or Federal Reserve Board or Federal Communications Commission is practicable, an independent, competent environmental bureaucracy can be developed. Simply making civil servants career employees, of course, is not sufficient. Government must create incentives to attract and educate competent, committed bureaucrats.

Educating American Business and Government

Public policy must also act to improve government–business communication subsequent to rule making. Government responsibility includes active education about the rationale for regulations, the approaches to achieving compliance, the enforcement policy, and available technical assistance. Educational campaigns should be substantive and specifically targeted.

In the past, the EPA had a notable technology transfer program, but it was cut back in the late 1970s, reportedly because of resource limitations. Education about regulation is a government function pursued much more fully in other countries. For example, in Sweden, the National Environment Protection Board spends a significant percentage of its budget disseminating information about its work. Administrators should give education divisions status equal to that of other operating units and provide funding at a level that allows detailed communication to affected businesses.

The costs of a full-scale education program about all aspects of regulation are considerable. But education makes the business firm aware that government notes its compliance record and that societal concern with environmental quality translates into direct contact by officials. Education promotes a sense of business involvement in policymaking, especially if government solicits evaluations of programs. A by-product may be more representative, more customized, and more detailed contribution to rule making than is achieved by traditional public participation mechanisms. Continuing discussion of the value and applicability of control choices can also result.

No doubt this contact with the regulated will at times be embarassing for government. Regulatory policy may demonstrate gross ignorance of the nature of a business production process, of the full costs of change, and even of the nature of a pollution problem. In fact, we offer this pressure on regulators as a benefit of greater government–business contact. Active education makes those who are responsible for regulation more accountable for the clarity and rationality of rules.

A complete education program involves interactions among government

---

[20] See also Representative Albert Gore's hearings on EPA, Sept., 1983.

and business leaders to assure that managers understand policy. Interactions should come at high levels and involve science advisors and environmental vice-presidents of the regulated firms. Education initiatives also might make use of roles analogous to those of agricultural extension agents to disseminate information about control technology at the plant level, focusing on those who are responsible for pollution control.

Advocating greater environmental education admits that ignorance of the minutiae of environmental law differs from ignorance of other criminal violations. Although those who advocate a strong environmental ethic may disagree, environmental noncompliance does not yet everywhere effect a sense of moral outrage. Statements by violators of their lack of knowledge of rules are often genuine, and the defense of "not knowing" the labyrinth of regulatory law can be convincing.

Education should also make clear the informational basis for controls. Science does not provide a rationale for some environmental standards and compliance attempts. In such cases, government enhances credibility and support for regulation in the long run (Roberts & Bluhm, 1981) if it makes explicit the value statement implicit in its regulatory policy. The regulator should make industry aware of the true rationale for controls; to suggest that a decision is based on scientific information when the science is unconvincing or skeletal is to give greater opportunity for charges of irrationality of the total regulatory program. Former EPA Administrator Douglas Costle has said that very weak data and a "grab a number" mentality characterized some early agency decisions. He explained that there were few health effects studies[21] available for recently developed compounds under regulatory consideration.

Enforcing Environmental Law

Our recommendations do not consider a strong enforcement policy as the top priority of efforts to decrease violations and increase the respectability of environmental regulation. On the other hand, to ignore reform in the enforcement process is to ignore an important policy lever of compliance policy, whether primarily command and control or fundamentally cooperation and consensus. Under any regulatory regime, certain classes of behavior will demand swift and efficient use of legal sanctions.

Communication of Enforcement Policy

Government should widely communicate enforcement policy but not enforcement strategy. *Policy* refers to statements about a commitment to environ-

---

[21] Interview, Mar. 1, 1982.

mental compliance and about the function of incentives and sanctions. Strategies are more specific; rather than inducing rule-obedient behavior, disseminating strategy-specific information can counter general compliance. For example, government should not inform business that a particular sort of hazardous waste litigation is a paramount agency priority. Nor need industry know which *mens rea* will be singled out for legal action. The enterprise should not be privy to whether the agency will make use of criminal sanctions or rely on civil law or less formal strategies or how the government will negotiate toward settlement agreements. Legitimate expectations of industry do not include probability statements about whether a government will prosecute, whether a plant will be inspected, the nature of monitoring activities, or the size of government caseloads. Business, however, can reasonably expect to know how a given administration intends to address environmental law enforcement. Messages about the importance of pollution control, coupled with other reforms noted in this chapter, can promote general compliance and greater respect for individual environmental protection initiatives.

Government might communicate the fact that it seeks to resolve environmental controversies without resort to the courts when business performance reflects good faith efforts to comply. The regulator should inform industry that it is not wedded to litigation and that its enforcement policy does not switch automatically to a suit when a possible violation is detected. Regulators can call on administrative process, on mediation, and on other strategies which emphasize information exchange (Dinkins, 1984). The government can demonstrate that it desires to reserve litigation for intractable firms and that consensual modes are preferable for many business–government disagreements. Just as enforcers should make it known that they will seek strict penalties in select circumstances, they should also clearly show the capacity to be flexible in enforcement. And regulators should demonstrate that the severity of penalties sought is based on overall government knowledge of a firm's performance including good faith efforts to comply.

Enforcement policy will demand respect if it also appears to be fairly applied, rather than based on firm size, government quotas, or an industry's vulnerability.

Sanctions

No one sanction is efficacious in all situations for realizing either specific or general compliance. Despite extensive commentary and opinion advocating a particular approach, only when government fits the nature of a penalty or incentive with the characteristics of the violating source will enforcement be effective. Although this notion of customizing sanctions is disfavored in the criminal justice system as applied to individuals, it remains useful when complex organizations are targets of compliance. Factors include the firm's cost–benefit calculations understood in terms of reputation and status as well as revenues and profits, its economic profile, and its organizational structure. Fines may be appropriate in one setting but not helpful in another—perhaps because *in terrorem* levels would seldom be imposed by the courts. The regulator should make it known that it is willing to seek severe penalties. Criminal sanctions may be the only reasonable response to Violation A but lead to Violator B's expensive countermeasures because of strong resistance to the criminal label. The firm's vigorous reaction may tax government's capacity to implement an overall environmental policy. And criminal sanctions may have no impact at all on standard operating behaviors of Violator C. This conclusion will be disappointing to those who seek universally applicable solutions to environmental problems. But the conclusion is inescapable in view of the evaluations of sanctions in most areas of law.[22]

In selecting sanctions, the agency should attempt to determine what appears to be motivating the business to violate regulations. Does a clearly identifiable businessman, like Distler, appear to be the source of a widespread problem? Does it appear that an industry-wide norm or practice is involved, or is the violation limited to a firm, as was the case in *Chrysler*? Is it likely that organized criminal elements are at work, as seems to be true in the New Jersey hazardous waste problem? Is the potential defense of ignorance of arcane regulations credible or do classes of violations seem to evolve from calculated risks? Other questions to be addressed: How well developed are the processes and procedures which result in noncompliance? Are they interconnected and subject to centralized control or difficult to classify, as in *Inmont*? Do violations occur regardless of who occupies important roles within the regulated business, or does a misfit between personnel and equipment or procedures appear to result in nonattainment of standards? Has there been a wanton disregard for health and safety? The more information the government has about categories of regulatory targets, the better can be its enforcement choices and the rationality and clarity of rules.

Prerequisites: Inspection, Monitoring

An effective enforcement policy requires an inspection and monitoring component which professionals carry out according to rules and schedules which only they know. Pollution sources should learn that government regularly watches activities with potential environmental impacts, but the regulator should not publish the timetable according to which monitoring occurs.

Inspectors should be well-paid and well-trained. They should rotate periodically among assignments to prevent the development of personal relationships

---

[22] See Chap. 5.

with the regulatee which can make inspection less vigorous. Admittedly, when inspectors rotate, they must relearn the regulatory setting, and this process has a cost in access to information. But using professionals will increase the probability that inspection is fully done and that review goes beyond *pro forma* compliance. Government should educate inspectors about the overall regulatory program and the rationale for rules so that they feel that they are part of a meaningful effort. People in the field should identify with policy and not act as if they, too, felt the imposition of standards to be unreasonable. As professionals they should have some latitude in deciding how a violation will be treated— whether it is *de minimis,* whether it requires a stern warning, whether it is inevitable in light of defects in the regulatory program. Enforcement agencies should not set quotas of citations for inspectors; rather, performance should be judged by long-range criteria of compliance.

Surprise should characterize the inspection schedule. Maintaining a degree of uncertainty about the time and the nature of visits will prevent business from coming into compliance only during the onsite review.

Covert monitoring will be necessary when there is a serious concern over immediate danger to public health. In the *Culligan* case,[23] which involved direct dumping of highly toxic materials into the Los Angeles sewer system, the city's hazardous waste strike force had to act as it did: at night it attached pollution monitoring equipment to the effluent pipes. The alternative was an absurd situation in which the city told Culligan about monitoring activity and the firm arranged to be in compliance only during inspections.

Surprise is not always easily managed. Fundamentally, government must balance surprise with legal rights, including the right to privacy and the right to proprietary information. When criminal charges are possible, the regulator must recognize criminal procedure requirements.[24] What is more, both sophisticated and incompetent business office practices can frustrate inspection. A receptionist can greet an inspector with time-consuming questions about the purpose of the visit or inquiries about the "firm which the gentleman represents." The company may treat a surprise inspection like any in a series of business calls throughout the day by salespeople. The receptionist may announce the inspector in the reception area, ask him or her to wait, and then call to find the proper host within the plant. Then more information may be requested. It may be some time before the inspection is actually made of production processes and plant equipment. Certainly, the firm will not easily cover major violations in these periods, but the delay can minimize the effect of surprise. Here, again, a professional inspectorate with industry experience will develop strategies—such as specific requests

[23] See Appendix for complete citation.
[24] See Note (Winter, 1982).

to see a production process or a particular employee immediately—that minimize obstructive responses.

Litigation Tactics

Finally, in those situations wherein compliance policy resorts to the courts, to be effective the regulator must incorporate intelligent litigation tactics. Management of the lawsuit is critical to effecting compliance.

Litigation competence translates to increased probability of compliance in several ways. Government thereby communicates to business that it can effectively pursue cases of alleged violations and overcome attempts to circumvent trials on the merits through procedural manuevers. Excellence in litigation carries a general message to the private sector: government treats environmental law seriously; although adversarial methods may not be the preferred means of promoting compliance, still, when initiated, they will be pursued vigorously and competently.

Effective litigation also promotes compliance in less direct ways. Litigation can clarify the nature of regulations and educate parties about the constraints on compliance. The latter benefit is often ignored in an antiregulatory environment that characterizes enforcement as a wasteful and hostile exercise. In fact, the trial on the merits is one of the most highly developed, and well organized means of channeling information across organizational barriers that arise—sometimes from other litigation—between the public and private sectors. Interrogatories, depositions, and informal exchanges among the parties can provide ideas on how to comply and even on new meanings of compliance.

Forum shopping may be valuable to effective litigation. The range of judicial philosophies and competence in environmental law is wide. Where possible, the prosecuting agency should litigate in courts with extensive experience with environmental regulation and seek judges who have displayed sympathy for the aims of a regulatory program. More practically, plaintiffs should avoid judges who have shown strong support for traditional, if not archaic, property rights. Some courts see the new environmental law as a means of problem solving and creative resolution of complex controversies; they will visit sites where violations have have allegedly occurred, ask incisive technical questions, and fully explore the bounds of new statutes. Others respond to environmental law much less favorably and are inclined to address cases on the narrowest possible ground (DiMento et al., 1980; Sax & DiMento, 1974). Enforcement staff with an air quality management district reported they recognized differences between the courts in neighboring counties and made their choice of forum appropriately.[25]

[25] Interview K.

Certain strategies and tactics of good legal practice apply to both government and business. Counsel's experts should be thoroughly prepared for trial. Otherwise, even a strong case can falter on understatement, overstatement, or irrelevant testimony. Lay witnesses should avoid testimony of environmental effects which will be viewed as narrow, parochial, or selfish (Frieden, 1979; Sax & DiMento, 1974). No matter what the final forum for litigation, clear articulation of the law is essential—perhaps more so that in other legal areas. Counsel will often have to educate judges about new and unfamiliar fields. For most judges, pollution control litigation represents a very small percentage of a caseload, and environmental law was not in the curriculum when they were students. Sidney Feinberg, of the United States Court of Appeals, noted that judges must be "led by the hand" through briefs in environmental law cases. He characterizes environmental law as "extraordinarily complicated," in part because environmental legislation is "endlessly long and opaque." He urges lawyers never to waive oral arguments in environmental cases.[26]

Assisting Support Groups

Stone (1975, p. 24) has noted that "part of the trick of changing the attitude manifested by the corporation as a whole is to locate the critical support group and strengthen its hand." Liroff (1980, p. 152) concluded about NEPA:

> Internal reorganization may promote compliance with NEPA, but only if it significantly alters communication flows and enhances access to the agency by supportive external sources. In short, it appears best to have a combination of internal and external change agents promoting compliance.

Assisting groups which focus pressure on target businesses and on enforcement agencies and which provide valuable information to government and the firm are themes common to this third set of reform suggestions. Common, too, is recognition of the need to maintain and coordinate support. The notion is straightforward: promote activity of those who put priority on compliance with well-made environmental law.

Continuing pressure on government agencies and on regulated firms to honor environmental law and to avoid violations is the most promising means of getting compliance. Support is found in *ad hoc* and established national groups of environmentalists, but it also exists within government and business. Policies that are aimed at increasing meaningful compliance should also aim to assist those with reasonable environmental agendas.

Recommendations to maintain the influence and to expand the roles of

[26] Judge Sidney Feinberg, Court of Appeals, Fifth Circuit, Remarks at Environmental Law Institute ALI–ABA Conference on Environmental Law, San Francisco, California, Feb., 1983.

support groups must recognize that some types of decentralized influence may be contrary to compliance goals. Policymakers face a choice. On the one hand, they welcome the assistance of those outside of government in fostering obedience to regulatory law. But they also recognize that these groups at times affect priority setting of the agencies, adding noise and confusion to the meaning of regulatory statements, that they may exaggerate the adversarial nature of government– business interactions, and that they can provide poor as well as good information for compliance decisions.

Support groups can focus influence, communicate environmental norms to business, and guarantee that government maintains its authoritative and legiti- mate function in the compliance process. In select situations, they can assist in establishing the order of compliance-seeking activity—enforcement, negotia- tion, and otherwise. And through public participation and litigation they can clarify the meaning of regulatory law. But these groups must themselves be disciplined; new information offered to guide or redirect the processes of effect- ing compliance must be subject to the same critical standards applied to informa- tion that other actors offer in all phases of environmental regulation.

Assisting support groups can be done in several ways: Provide channels for their messages about environmental quality, for example, through greater access to the legislative function and regulation writing. Remove obstacles to support group participation in enforcement. Reward and otherwise encourage and protect those both within and outside the firm who promote compliance.

Access to the Legislative Function and to Regulation

We have discussed earlier in this chapter the functions of interest groups in the processes of scoping environmental policy. Here we advocate increased use of support groups in the early phases of legislation and of rule making.

Much of what we advocate is already sought as an ideal in the legislative process or found in bits and pieces in the procedural guidelines of administrative agencies. But relatively minor reforms can remove some obstacles to making and implementing consistent environmental policy.

Government should provide incentives for support groups to participate in legislative hearings and in rule-making proceedings, including negotiated rule making. These incentives must go beyond simple publication of schedules for testimony taking. One form is to provide monetary compensation for representa- tives of groups with opposing views on proposed rules. A version of the Magnu- sson–Moss[27] model may be appropriate. Support groups identify spokespersons for their positions. Government decides how many versions of a position will be

---

[27] 15 U.S.C. § 57a(h)1 (West Supp. 1976).

represented. Interest groups thus contribute to environmental policy as it is made. Although the full realization of this idea creates some front-end costs, it avoids many aspects of the wasteful challenges to poorly formed, nonrepresentative, or unrealistic law. And it provides the financial wherewithal for those with strong interests in the compliance process to pass information to policymakers.

We emphasize that involvement must remain subject to the direction and control of government—in most cases, administrative agencies. We recognize that tension exists between professional centralized control of environmental policy and decentralized participation.

The question of the proper role of citizens in environmental law is most pointedly raised about the function of the initiative. Citizens can in many jurisdictions independently make law, preempt legislative activity, and through the referendum negate legislation.[28] Some state constitutions explicitly provide that all legislative authority is lodged in the citizenry: the legislature is allowed by the sovereign body politic to make law. This general principle of citizen sovereignty can be implemented in various ways. Should further use of the initiative be advocated if compliance is the public policy aim? We offer a qualified yes to this question and at the same time recognize some constraints on the initiative.

When properly channeled, the initiative improves both the communication of and support for environmental law. The initiative relies on information held by large numbers of people who directly experience a problem. It has a mixed history; critics question both its wisdom and its efficacy. Opponents contend that laws so passed paint with too broad a brush, are based on widely held but limited information, and do not benefit from the learning that takes place in legislative hearings and administrative lawmaking proceedings. Use of the initiative as a general environmental policy aimed at fostering compliance must be selective; citizen-based lawmaking is not appropriate to address all pollution control and natural resource management objectives. Indeed, there have been numerous initiatives which were technically deficient or reflective only of short-term public concerns.[29] The initiative should be employed when those to be regulated legislatively block a worthwhile environmental policy or when noncompliance results from the ability of interest groups to dilute radically environmental rules.

When properly drafted, an initiative is a clear, noncompromised, and more specific statement of a societal aim than ordinarily is the product of legislative process and interest group politics. Indeed, the message of an initiative can be so instrumental as to fail to enlist broad understanding of its purpose. Therefore,

[28] Seventeen states allow for the initiative in their constitutions. Council of State Governments, *The Book of States* (1982).

[29] See Institute for Governmental Studies, California Data Brief, "Propositions: The Initiative In Perspective," University of California, Berkeley (Apr., 1982).

when employed, the initiative should introduce to the electorate the rationale for new law and the meaning of specific provisions. The initiative then serves an educative as well as a lawmaking function.

The case of the California Coastal Zone Conservation Act[30] represents a successful—albeit controversial—use of the initiative in environmental law. In the 1970s three legislative attempts were made in the state to address the need to protect the ecologically sensitive seawaters under the control of the state and the areas between the ocean and the coastal foothills. This expanse of coastline is among the most beautiful and valuable in the world; and means were sought to insure its wise management and protection. Each bill in the California legislature failed in part because of strong opposition of those who would have to comply, including the real estate industry, land companies, builders, and local government.

In 1972, the public succeeded in having an initiative placed on the ballot; it passed with 55% of the vote. The initiative, which was extremely unpopular with many of those it regulated, had several compliance-promoting attributes. The mandate was clearly articulated and its general rationale was well communicated in its preface: "To preserve, protect, and where possible, to restore the resources of the coastal zone for the enjoyment of the current and succeeding generations. . . . The permanent protection of the remaining natural and scenic resources is a paramount concern to present and future residents of the state and the nation." Representatives on the Coastal Commission, composed of six regional bodies and a state commission, would have a commitment to the central aim of the law, that is, careful scrutiny of decisions on the marine environment, esthetics, development, and recreation on the California coast. Although enforcement resources were not great, the mandated permit process made the law action-forcing because no construction could proceed in the coastal zone without a permit. The government–target communication channel thus was direct. Opportunities for easing the regulatory impact of the law were limited; to amend it a new initiative was required. Finally, the commissions were single-purpose bodies which could concentrate their activities on coastal management.

Access to Judicial Review of Environmental Violations

Support groups should not depend on litigation to channel influence and pressure on government and industry. It is unrealistic, however, to suppose that environmental constituencies can be effective without at least the threat of lawsuits to back up their sources of influence. Access to the courts can assist groups who promote compliance with environmental regulations.

---

[30] California Coastal Zone Conservation Act, 18 Cal Pub. Resources Code §§ 27000 *et seq.* It became effective Nov. 8, 1972.

Citizen suit provisions are found in statutes such as the Clean Air Act and Federal Water Pollution Control Act Amendments[31] or as general substantive environmental rights provided for in the public trust doctrine and legislated in natural resources and environmental protection acts.[32]

Extensive arguments for and against the citizen environmental lawsuit have been made in law review and other commentary (Bryden, 1976; DiMento, 1977; Sax & Conner, 1972). Fundamentally, the debate turns on evaluations of judicial competence, the opportunity costs of litigation, and predictions about the responsibility of private citizens when their legal powers to promote compliance are enhanced. Analyses of the effectiveness of the citizen suit reach different conclusions.[33] Nevertheless, when employed selectively the citizen suit allows legitimate support groups to oversee the enforcement activities of agencies and to influence directly corporate and public agency response to environmental concerns. *Ad hoc* environmental groups, large environmental organizations such as the Sierra Club, public interest law firms from the Pacific Legal Foundation to EDF, private individuals, and businesses themselves have direct access to state courts and the federal judiciary to seek review of allegedly noncompliant activity. This citizen-initiated push has been required to implement a significant percentage of environmental laws.

With adoption of reforms that promote operational law, that advocate consensual approaches to compliance whenever possible, that recognize the need for a senior-level civil service and for rigorous enforcement in circumscribed situations, the citizen suit nonetheless remains important.

Selective use of citizen suits (the laws are sometimes referred to as private attorney general provisions) can supplement insufficient enforcement resources in state and federal agencies. Citizen suits allow challenges to organizational decisions that limit enforcement. They subject environmental agencies to public scrutiny regarding the rationality and continuity of environmental policies and the clarity and consistency of environmental regulations. They directly and authoritatively inform business that support exists for environmental objectives (DiMento, 1977). Properly managed, they are a responsible channel of support group influence.

Citizen suit provisions need not be open-ended. They can be drafted to address the legitimate concern that litigation jeopardizes realization of priority goals and that citizens will file frivolous and harrassing suits. Litigation can be limited to situations in which government has failed to develop or enforce environmental rules in a timely manner. Also, statutes can exempt specified activities with special economic or other significance from citizen suits, for exam-

---

[31] 42 U.S.C. §§ 7401–7642, 7604; 33 U.S.C. §§ 1251–1376, 1365.

[32] An example is MEPA. See Chap. 3, p. 48.

[33] Compare Sax and DiMento (1974) and Haynes (1976) and Bryden (1976).

ple, experiments with new forms of regulation. Security bonds can be set at a level to cover potential costs of delay to challenged projects when the defendant ultimately prevails; and clear definitions of compliance standards can be drafted. Such restrictions can act to screen litigation that would obstruct professional agency efforts to achieve compliance. Similarly, joinder of plaintiffs can limit litigation while promoting coordination of compliance efforts.

Channeling Support, and Support Within the Firm

American business is highly heterogenous in its support for environmental control. Employees who actively favor stricter emission controls, more careful analysis of ultimate despositories for wastes, and more national preservationist policies work side-by-side in industry with employees who consider environmental regulation to be anti-American. The range of environmental sentiments within American business is broad.

The spectrum of corporate organizational responses to environmental law is similarly wide. Many American companies have offices of compliance, and some are models of responsiveness to law. Nonetheless, many small businesses have not recognized the importance of institutionalizing compliance. And several large firms design the compliance function in narrow, sometimes counterproductive and begrudging ways. An effective compliance team does not simply react; it monitors the political environment with attention to regulation. It establishes means to articulate the firm's perspective; it advises on proposed rules and legal appeals; and it develops compliance systems. Business and government must recognize more fully sincere efforts within industry to meet the compliance function. If compliance is to be treated seriously, those divisions and persons who promote compliance must become more central to management.

One means of rewarding support within American business for intelligent environmental regulation is through appointment to agency advisory committees. Business participation can promote rationality of regulations and receptivity to rules. Choices should be business people committed to reasonable interpretations of the aims of environmental law. Advice from committees of outliers within interest groups will differ greatly from that generated by those with more normative positions within a collectivity. Advisory committee service can be considered a stage in lawmaking and not a political reward to people who have some opinion on environmental law, no matter how extreme. Advisory committees are useful to the extent that they offer and review regulations, evaluate the law in action, and make recommendations for change.

Within the firm, recognition can be given to those responsible for seeking innovative, cost-effective approaches to compliance. And companies can provide increased influence and status to environmental executives—for example, by allowing line access to the firm's CEO and chief operating officer. Companies can reward compliance divisions according to their performance in meeting environmental standards, and environmental groups can look to the private sector for candidates for environmental awards.

A source of considerable information about corporate compliance is the firm's employees. Both societal and corporate benefits can derive from knowledgeable employees reporting violations or potential violations of environmental law. Presently, however, several factors inhibit the individual from acting as societal guardian or as whistle blower. The most obvious fear is loss of one's job. But more subtle and culturally pervasive inhibitors exist. The whistle blower has been seen as disloyal, a traitor, a person who aims to promote disunity within the company.

Three reform routes are recommended to assist the whistle blower. The business sector can make available internal offices, separate from central management, for disclosure of noncomplying activities. A type of ombudsman might give workers an opportunity to disclose observed problems secretly and to be rewarded for valid information. The ombudsman, who must be an autonomous, independent member of the corporation, would then decide which complaints would be carried to management. The employee could assess subsequent corporate response before deciding whether to report information to regulatory agencies. Second, rules should guarantee greater worker protection, as is the case in Michigan for whistle blowers (Braithwaite, 1984). Third, a cause of action for malicious dismissal should be legislated (Stone, 1975).

A Final Word

There is no magic way to bring about compliance. In the last two decades, environmental law has become immensely complex. Analyses of the problems of environmental degradation have proliferated. There has been some convergence among business and government observers but there remains considerable disagreement over problem identification. Meanwhile violations, some only irritating and some egregious, have continued—indeed, have increased in number. There is no single lever that government agencies can pull, no single strategy that American business can promote, which will result in compliance with our pollution control and environmental protection laws.

To achieve compliance, public policy must recognize the variety of participants involved in the compliance event, the range of their motivations, and the individual, small-group, and organizational dynamics at play. To view the compliance problem in this way is to see that a number of reforms are necessary— happily none too drastic or radical.

We have presented in this volume a considerable number of suggestions. Not all of them will apply to the analysis of any one environmental law. Indeed, some are incompatible when focused on a single problem. This conclusion should not be surprising with the recognition that threats to the environment come from the actions of well-meaning but incompetent businesspeople and government officials; from the environmental criminal's direct assault on nature; from standard operating procedures of giant organizations and from agencies with benign plans; and from American industries which conclude that what they do, no matter what the effects on the air, water, and land resources, is good for the country.

Multiple strategies for gaining compliance also recognize the wide range of local and national support for individual environmental laws. On one side of a continuum are goals expressed in sections of the Clean Air Act and in revered local legislation protecting ecologically sensitive areas. Promoting compliance with these laws poses a challenge much different from that of implementing, for example, the Noise Control Act, whose constituency has never been fully identified, whose rationale has not been well communicated, and whose demands, consequently, have fallen into relative legal oblivion.

We propose a policy goal more modest than some analyses of regulatory reform. That aim is to eliminate egregious, serious, and willful violations of law that translate to environmental harm and to move in the direction of overall objectives of environmental law as developed by serious, committed persons within and outside government.

Our modest approach will be compared with advocacy of radical surgery on corporate America and business–government relations. Whatever route is chosen (or, more probably, approximated) must be evaluated by an answer to the fundamental question about inducing compliance: Have government and industry worked to decrease exposures to environmental pollutants and to control degradation of environmental quality?

APPENDIX

List of Cases

Adamo Wrecking Co. v. United States, 545 F.2d 1 (1976), 434 U.S. 275 (1978)

The United States Supreme Court reversed a court of appeals decision and held that the EPA's authority to set emission standards did not include authority to set a design or operational standard. Emission standards were read to be quantitative emission limitations, not work practices. The EPA administrator cannot make a regulation an emission standard by "mere designation." The issue arose in a criminal enforcement in which Adamo Wrecking Company challenged a rule which required that before demolishing old buildings the wrecker must wet down asbestos insulation. Adamo had run afoul of the regulation in the demolition of a four and one-half story commercial masonry building in Detroit. "When is an emission standard not an emission standard?" was a particularly important question because Congress attached stringent sanctions to violations of emission standards—sanctions of a kind that did not apply to other administrative orders. Also, the promulgation of an emission standard was not subject to judicial review in criminal or civil enforcement actions.

Congress subsequently reversed the Adamo holding, which may have reflected the high court's concern over limitations in the Clean Air Act on judicial review of air quality standards.

People of the State of California v. Aaro, Inc., C 411941 and 31227062, Los Angeles Municipal Court (1982)

A March, 1982, inspection of this Los Angeles battery recycling firm revealed that hazardous waste was being drained directly from tanks into the city sewer system. Aaro could produce no manifest detailing the procedures it followed in handling hazardous wastes. Because of the firm's history of neglect of warnings and citations, the city of Los Angeles sought to obtain evidence which would support criminal sanctions. On May 13, 1982, and following secret electronic monitoring, a raid uncovered allegedly illegal dumping of a highly acidic material. In Los Angeles Municipal Court, the city subsequently filed both civil and criminal proceedings against Aaro, its president, and a foreman for violations of the California Health and Safety Code and the Los Angeles Municipal Code. The city sought jail terms for the defendants because of the severity of the violations and the prior history of the firm. Aaro's sewer line was immediately capped. The defendant pleaded *nolo contendere* and was sentenced to three months in jail and 36 months on probation.

People of the State of California v. Aerojet General Corp. and Cordova Chemical Co., 286073 (1979)

Aerojet General's rocket fuel manufacturing plant is located on an 8,000 acre tract in Sacramento County, California. In August, 1979, nearby residents were warned that their drinking water, obtained from local wells, was contaminated with trichloroethylene (TCE), a suspected carcinogen. State tests conducted in September, 1979, traced numerous toxic and hazardous chemicals to water coming from Aerojet. A civil lawsuit filed in December, 1979, accused Aerojet of violations of the State Water Code in dumping chemicals into surrounding groundwater through unlined ponds and swamps since 1963. The suit sought penalties of $6,000 per day from all defendants. Aerojet had allegedly dumped up to 20,000 gallons of hazardous discharges per day into the groundwater, "in reckless disregard of public health." Aerojet denied the charges.

People of the State of California v. A–Z Decasing, C 405741 (1982)

On June 14, 1982, after soil samples taken both in and near the plant indicated lead levels up to twenty times the legal limit, the Pomona, California, battery recycling firm was ordered to contain lead and sulfuric acid-bearing dust particles on its property and to protect its employees from harmful exposure by vacuuming the dust. Although civil penalties of $2,500 for each violation of both the California Health and Safety Code and the Business and Professions Code might have been sought, in September, 1982, a Los Angeles Superior Court refused to take any further action. The judge apparently was satisfied that problem areas had been corrected and was impressed with the firm's history of supporting stringent cleanup requirements.

People of the State of California v. Bob's Plating, 31253286, Los Angeles Municipal Court (1983)

Following the Los Angeles firm's long history of noncompliance with regulations governing chemical dumping, in March, 1983, the Los Angeles Hazardous Waste Strike

Force secretly monitored Bob's Plating's discharges into the city sewer system. Inspectors discovered several violations, including excessive amounts of copper, zinc, chrome, and cyanide as well as low pH levels (low pH and high levels of free cyanide can result in the release of hydrocyanic acid, the gas used in prison gas chambers). On March 31, 1983, Bob's Plating was thoroughly inspected and shut down. Inspectors had found "a total lack of housekeeping" and 3–4 inches of sludge on the floor of the facility. Hearings before the Los Angeles Board of Public Works during April and May of 1983 resulted in revocation of the company's discharge permit. The plant was to remain closed until Bob's Plating met all regulations, requiring the firm to purchase new plating and pollution control equipment. The company's general manager was sentenced to and served four months in county jail.

People of the State of California v. Capri Pumping Service, Los Angeles Superior Court (1981)

In June, 1980, the Los Angeles Superior Court ordered Capri Pumping Service, a recycler of hazardous wastes, to cease storing hazardous wastes and contaminated soils. For several years, Capri had allegedly abused waste processing regulations under the California Aministrative Code. Capri's owner was found in contempt of court in December, 1981, for storing such wastes in violation of the above order. He was fined $3,000 and sentenced to thirty days' imprisonment. This sentence was suspended in February, 1982, in lieu of the company's agreement to clean up the site within six months. The contempt charges were later purged by the defendant's compliance with a court order, and the judge concluded that "the time for any danger to the community has long since passed."

People of the State of California v. Culligan, 31242629, Los Angeles Municipal Court (1982)

An investigation by the city's Hazardous Waste Strike Force indicated that over a period of eight months, Culligan Deionized Water Service had dumped from six to twelve tons of hazardous waste directly into the city sewers. Culligan, which had never had a permit to process the waste, admittedly had handled the heavy metal wastes since 1976. On December 1, 1982, the city attorney filed 73 criminal charges in Los Angeles Superior Court against Culligan. The Board of Public Works made its final decision to revoke Culligan's industrial permits on January 25, 1983, and imposed a $205,000 penalty on the company. In the January hearing, a former Culligan sales manager testified that "there was no attempt" to control the discharges. Culligan's president was subsequently sentenced to serve ninety days in jail because of the violations, and an additional $100,000 fine was levied against the firm. In addition, restitution was required in the form of free distilled water and purified water service to nonprofit hospitals. In July, 1983, the original administrative fine was withdrawn.

People of the State of California v. Custom Plating Corp., C 415339 (1982)

After receiving fines of $250 and $500 in 1980 for criminal misdemeanor convictions, Custom Plating Corporation agreed to a preliminary injunction on June 22, 1982, which prohibited the firm from dumping cyanide, copper, chromium, nickel, and zinc into the Los Angeles sewers. The civil suit charged that Custom Plating had dumped excessive levels of the substances into the sewers, violating the California Health and Safety Code, the Business and Professions Code, and a Sanitation District wastewater ordinance. In addition to requiring the firm to maintain and improve neutralizing equipment, the state sought civil penalties in the form of fines. Custom Plating denied having exceeded any wastewater discharge limits.

People of the State of California v. Garrett Corp., C 416199 (1981)

On August 12, 1981, a forklift operator for the Air Research Industrial Division of Garrett Corporation mistakenly dumped several 55-gallon drums containing hazardous oil and metal filings into a trash bin in the LaPuente, California, landfill. A civil suit filed by the state charged a violation of the California Health and Safety Code, namely, depositing hazardous wastes in a location unsuited for that purpose. A negotiated settlement was worked out in Los Angeles Superior Court in July, 1982, and Garrett agreed to pay a $5,000 penalty.

People of the State of California v. Inmont, C 365979 (1981)

For more than ten years public agencies attempted to compel cleanup and disposal of 14,000 drums of improperly stored hazardous waste at this hazardous waste storage facility in Santa Fe Springs, California. After it became clear that there were no public funds readily available for the needed cleanup, various agencies involved agreed to file a civil lawsuit against the current and previous owner of the site, and against six paint companies believed to have been responsible for materials on the site. The state sought permanent and injunctive relief. If Inmont could not finance an adequate cleanup, then the state would go ahead with court action. However, on July 10, 1981, an arsonist set fire to the site; the inferno lasted five days. Ultimately, a settlement was reached: Superfund would pay $1 million, Inmont $1 million, and the city $250,000. The cleanup was completed as of January, 1982.

People of the State of California v. Narmco, M 99196 (1980)

On April 4, 1980, an inspector for the South Coast Air Quality Management District found Narmco, a Costa Mesa, California, plastics producer, in violation of a section of the California Health and Safety Code. Narmco had allegedly created a public nuisance, as neighbors' complaints had alerted authorities to the firm's discharges of smoke, odors, fumes, and gases. On February 21, 1981, in Orange County Harbor Judicial District Court, Narmco pleaded *nolo contendere* to the charges of releasing the offending emissions. Narmco's attorney stated that even though the company was innocent, it was "not worth the cost involved to prove it." Narmco paid the maximum $500 fine, and the firm was placed on two years' probation.

People of the State of California v. Southern California Edison, C 372982 Los Angeles Superior Court (1981)

In July, 1980, after the rupture of a Southern California Edison (SCE) power pole-mounted transformer, homeowners whose backyards were contaminated with the carcinogen polychlorinated biphenyl (PCB) sued the power company. The Los Angeles County District Attorney's Office conducted a subsequent investigation and revealed that Edison's procedures for cleaning up such spills did not meet federal or state requirements. A complaint was filed in Superior Court for an injunction and civil penalties against SCE for failure to clean up the spill properly. In addition, the prosecution alleged violations of the California Business and Professions Code so that the amount of the penalty could be increased. The final judgment ordered SCE to pay a fine of $61,000 and to pay court costs of $24,000; furthermore, SCE was ordered to follow federal guidelines in the cleanup of future spills, to apply for all permits required when handling PCBs, and to follow specific guidelines for reporting such spills.

Chrysler Corporation v. EPA, 627 F.2d 1095 (1979)

In the first Clean Air Act automobile recall for design and maintenance rather than for manufacturing defects, EPA ordered Chrysler Corporation to recall 208,000 1975 Chrysler, Dodge, and Plymouth automobiles. Performance tests indicated that under actual driving conditions improper design and adjustment of carburetors led to emissions of carbon monoxide above the legally allowable fifteen grams per mile. EPA argued that Chrysler's emission control system was so complex that ongoing compliance could not possibly be realized. Chrysler challenged the recall, charging that it should not be held responsible for the actions of private individuals whose task it was to adjust a carburetor idle so that CO standards were met. In September, 1979, the United States Court of Appeals upheld the recall. The Supreme Court denied petitions for writs of *certiorari* in December, 1980.

Florida v. Hooker (1981)

A unit of Hooker Chemical Corporation, the Occidental Chemical Company, agreed to a $1.1 million settlement in a suit brought by the Florida Department of Environmental Regulation concerning excessive emissions from a White Springs, Florida, fertilizer operation. An internal review by the parent company in 1977 discovered that employees had altered the manufacturing process during semiannual tests for emissions so that the plant would meet state standards. The company released the employees thought to be responsible for the scheme.

Fullerton Homeowners v. McColl et al. (1983) Orange County Superior Court, Lead Case Nos. 35-70-39 (Gorber et al. v. Shea et al.) and 37-48-02 (Cosman et al. v. Chevron et al.)

This seven-acre site in Fullerton, California, was used as a wartime dumping ground for 100,000 tons of toxic refinery waste from 1942 to 1946. Although in 1977 the city learned of dangerous levels of fumes emitted from the site, not until September, 1980, did state health officials notify almost 900 residents living within two thousand feet of the dump that emissions of SO_2 posed a significant health hazard. Subsequently, in Orange County Superior Court, 65 families sued McColl and several land development firms and oil companies for careless conduct in allowing the area to become a health hazard. Other defendants were Chevron USA, Inc., Chevron Land Development Co., Standard Oil of California, Standard Oil USA, Fullerton Hills Development Co., ARCO, Union Oil, Texas Oil, and the William Lyon Co. The plaintiffs seek general and punitive damages, attorney and court costs, and medical costs aggregating to some $100 million. The state is studying the dump in hopes of devising a feasible clean-up plan. But until the impacts of the waste are better understood, transportation of the wastes has been enjoined. All studying the dump in hopes of devising a feasible clean-up plan. But until the impacts of the waste are better understood, transportation of the wastes has been enjoined. All "McCall Dump" cases were consolidated and in November, 1985, a mandatory settlement conference was scheduled.

General Motors Corp. v. EPA, 80-1868, 80-2027, 81-1029

In June, 1980, EPA ordered General Motors to recall 169,000 1975, 1976, and 1977 Cadillacs and Oldsmobiles because faulty exhaust systems prevented the automobiles from meeting federal antipollution standards. In November, 1980, GM filed suit in the United States Court of Appeals challenging both the EPA's recall and the May EPA Interpretive Rule which made the recall possible. GM's challenge concerned action involving 1975 Cadillacs. The corporation felt that, under the Clean Air Act, EPA had no authority to recall automobiles which had exceeded their "useful lives"—5 years/50,000 miles. The EPA contended that the act requires vehicles to comply regardless of their age or mileage. In December, 1983, a federal court of appeals ruled that the EPA had violated the Clean Air Act by ordering the recall. But the court reversed its ruling in September, 1984, and held that the manufacturer must repair pollution-related defects without cost if they are discovered during the vehicle's "useful life," even if the repairs are made much later.

Hooker Chemical Co., Ruco Division, v. EPA, 642 F.2d. 48 (3d Cir. Feb 23, 1981)

On several occasions, beginning in January, 1977, and until May, 1979, Hooker Chemical Company released vinyl chloride from relief values in a plant in Burlington,

New Jersey. Hooker reported these releases to EPA. EPA directed Hooker Chemical Company to prevent future releases of vinyl chloride from its industrial facilities. Later the EPA withdrew its compliance order. Hooker challenged the legality of the compliance orders, arguing that at the time of the order the EPA lacked authority under the Clean Air Act to direct specific work practices as part of emission standards. The Third Circuit Court of Appeals declined to rule on the legality of the orders, charging that the issues were not ripe for review. The EPA was continuing its investigation of the discharges and had not ordered any specific compliance measures.

Kennecott Copper Corp. v. EPA, 526 F.2d 1143; 424 F.Supp. 1217 (1976); 572 F.2d 1349 (1978); 684 F.2d 1007 (1982)

The battle between Kennecott and the EPA over acceptable SO$_2$ emission controls began when Kennecott sought judicial review of an EPA-promulgated provision of the Nevada state implementation plan (SIP). Kennecott opposed the EPA regulation which required control of SO$_2$ emissions at the McGill, Nevada, smelter by means of an acid plant. In *Kennecott I*, the EPA regulation was upheld. Subsequently, Kennecott closed its plant. The company next petitioned the state of Nevada to revise its SIP to allow alternatives to the acid plant, specifically curtailments in production and dispersion by means of tall stacks. Nevada approved, and Kennecott then filed suit in district court arguing that the administrator of EPA had a mandatory duty to accept the revision. The district court ruled in favor of Kennecott, but the United States Court of Appeals set aside the lower court's preliminary injunction requiring approval. The court of appeals maintained that the administrator was not bound to accept the state's conclusion about economic infeasability of the original EPA rules and that judicial intervention was improper until EPA had actually acted on the SIP revision. The court remanded the case to the district court, with instructions to enter an order dismissing the action. In the 1982 opinion, the United States Court of Appeals for the District of Columbia Circuit concluded that the EPA overgeneralized the holding of the earlier court of appeals opinion. The court vacated EPA's regulations for nonferrous smelters, stating that Congress ''sought to provide relief to the nonferrous smelter industry from a number of adverse economic consequences short of actual shutdown'' (p. 1014).

North Carolina v. Burns and United States v. Burns, 512 F.Supp 916 (1981)

Under the direction of the defendant, the defendant's two sons allegedly dumped PCB-laden transformer oil in a four-to-six-inch band along two hundred miles of rural North Carolina roads in 1978. Prosecuting attorneys characterized Burns's behavior as going beyond mere disobedience; it was ''laced with arrogance.'' Defense attorneys argued that their client should not be harshly sentenced because he had not committed a crime; he had only broken a governmental regulation. As part of a plea bargain to keep his sons out of prison, Burns agreed to testify against his employer and pleaded guilty in Halifax Superior Court of violating the Toxic Substance Control Act of 1976. The state court sentenced him to serve from three to five years in North Carolina prisons. When released, Burns is to devote eight hours per week to environment-related public service activities. (In the United States District Court Burns was sentenced to serve 18 months in federal prison.)

Pennsylvania v. William Lavelle et al., 83 CR.615(a) and (b), 83 CR 673, 83 CR 674

In Dauphin County, Pennsylvania, a grand jury indicted Lavelle and his companies, W. A. Lavelle and Son and Lavco, Inc., for violation of Section 911 of the Crimes Code of Pennsylvania. The August, 1982, racketeering charges stem from the defendant's alleged deception of firms with which the defendant contracted for the disposal of hazardous wastes. The case was the first ever to employ antiracketeering laws for toxic dumping. Between 1976 and 1979, Lavelle allegedly illegally buried 9,721 drums of waste and dumped more than 3.5 million gallons of bulk liquid industrial wastes into a garage floor drain which fed into an abandoned anthracite mine beneath Scranton. Lavelle had secured contracts which stated that the waste would be disposed of in compliance with law. The firm allegedly earned more than $500,000 through the "series of thefts by deception." Lavelle was convicted in April, 1984, in a nonjury trial.

Pruitt v. Allied Chemical Corp., 523 F.Supp 975 (1976)

A subsidiary of Allied Chemical Corporation allegedly spilled a large quantity of Kepone, an extremely toxic pesticide, into the James River. The spill eventually reached the rich fishing area of the Chesapeake Bay, causing an enormous environmental disaster. A coalition of commercial fishermen; seafood retailers and wholesalers; restaurateurs; and marine, tackle, and boat and bait shop owners alleged that the spill had adversely affected their economic interests. The lawsuit was based on nuisance, negligence, and the Federal Water Pollution Act. A United States District Court employed a "direct–indirect" test to determine which of the parties had standing. Only those who have an intimate and fundamental relationship with the natural resource can rightly claim standing; on September 15, 1981, the court ruled that only the commercial and sports fisherman had the necessary relationship upon which to base a claim for negligence.

SCAQMD v. Powerine, SEC. 40140 (1983)

For nine days during July, 1982, the Santa Fe Springs, California, oil company released a "rotten-egg" odor from its refinery. After the odor had elicited 200 complaints from neighboring residents, the Norwalk Superior Court gave the company 72 hours to solve the problem. The company complied. On August 11, 1982, following citations for releasing corrosive wastes, for not applying for an industrial waste permit, and for releasing excess levels of dissolved sulfides, final notice was served for discharges into a sanitary sewer. However, the company was not shut down, and none of the citations has led to the assessment of any penalties. The company has apparently been trying to work out a settlement with the South Coast Air Quality Management District. By July, 1983, however, an action had been filed in Superior Court concerning thirty-four rule violations.

United States v. Derewal et al., 77-287-1, United States District Court, Eastern District of Pennsylvania (1978)

On March 29, 1977, Philadelphia police observed several men apparently discharging the contents of a chemical tanker trailer rig into the Delaware River. None of the men could produce documents detailing the contents of the trailer. The men were arrested. Subsequent investigation of the businesses associated with the men—Environmental Chemical Control and Derewal Chemical Company—yielded evidence that the companies had been dumping unknown amounts of chemical and industrial wastes directly into the sewer system. In July, 1977, a federal grand jury charged the defendants and their companies with violations of the Refuse Act, the Federal Water Pollution Control Act, and the Clean Water Act. The defendants entered guilty pleas in United States District Court on June 19, 1978. Manfred Derewal, president of Derewal Chemical, was sentenced to six months' imprisonment, fined $20,000, and placed on probation for four and one-half years.

United States v. Distler, 671 F.2d 954; 454 U.S. 827

Two chemicals were found to be responsible for the harsh smell which irritated workers and forced the closing of Louisville's main wastewater treatment plant in March, 1977. The two chemicals—called hexa and octa, for short—were dumped into Louisville's sewers around March 1. Subsequent investigations traced the chemicals to the Kentucky Liquid Recycling Corporation, and to several of the firm's employees. In December, 1978, the firm's president, Donald Distler, was convicted in United States District Court on two pollution charges. Other members of the firm had been indicted but were not convicted. Distler was sentenced to two years' imprisonment and a $50,000 fine—the toughest criminal penalty ever handed down in a federal pollution case. Distler appealed the decision to the United States Supreme Court, but the decision was upheld.

United States v. Frezzo Brothers, Inc., 642 F.2d.59 (1981), 703 F.2d62

The Frezzo brothers, who own and operate a mushroom business in eastern Pennsylvania, allowed water which had circulated with manure to escape from a concrete storage tank. The runoff eventually drained into the White Clay Creek. In 1977 and 1978 a County Health Department investigator collected samples of runoff from the Frezzo establishment. In 1978, a jury in district court found the brothers guilty of negligently discharging pollutants in violation of the Federal Water Pollution Act Amendments of 1972. On appeal, the United States Court of Appeals, Third District, held that the runoff was not from a point source and thus did not require a permit, assuming the pollution was agricultural. The lower court's decision was reversed and remanded. Nonetheless, in 1983 the same court of appeals held that the Frezzo operation was manufacturing, not agricultural; thus, the Frezzos were still subject to criminal sanctions.

United States v. Gleaton, 80-00030-A, United States District Court, Eastern District of Virginia (1980)

Marion A. Gleaton operates a sewage treatment plant in eastern Virginia. On February 8, 1980, the United States Attorney charged in United States District Court that Gleaton had failed to retain possession of records detailing the monitoring of his sewage plant, had willfully and knowingly falsified monitoring readings, and had entered false statements into records thirty-four times between December, 1976, and September, 1979. Gleaton plea bargained and pleaded guilty to one count of entering false statements and to falsifying monitoring readings in violation of a NPDES permit. The remaining charges were subsequently dropped. Gleaton was sentenced to six months in jail, fined $2,000, and placed on two years' probation. The jail term was suspended.

United States v. Hooker and New York v. Hooker (1980) CIV-79-987C, CIV-79-988C, CIV-79-989C, CIV-79-990C, and ELR 20801

Until the 1940s, the Hooker Chemicals and Plastic Corporation dumped about 22,000 tons of chemical waste into a fifteen-acre area around the Love Canal in Niagara Falls, New York. The canal itself is an abandoned hydroelectric trench about 40 feet deep. The Niagara Falls Board of Education purchased the canal site from Hooker in 1953 for one dollar and built on it a playground and school. Homes were also constructed in the Love Canal area. Recent investigations have detected more than 300 chemicals in the vicinity, including ten known or suspected carcinogens. Extremely high levels of dioxin have been detected. Over the years residents have suffered from a variety of health problems ranging from cancer to hyperactivity. In 1977 chemicals appeared on the surface and in basements of homes in the area. Residents were then relocated and cleanup operations began. The United States sued Hooker for clean up costs. In January 1980, the state of New York filed a $635 million lawsuit against Hooker and its parent company, Occidental Petroleum; the state charged that the company failed to prevent chemicals from migrating out of Love Canal, thereby exposing nearby residents to grave danger. Furthermore, the company allegedly failed to warn residents of this danger.

Settlements have been reached in several of the proceedings involving Hooker. They address, among other activities, means of containing migration of chemical wastes, protection of drinking water, use of remedial technology at the site, and insurance against nonperformance.

In a related matter, Occidental, the Board of Education and the City of Niagara Falls reached an out of court settlement with individual claimants who sought damages of $16 billion. Approximately 1,300 former residents of the area received awards ranging from $2,000 to $400,00 from a fund to which Occidental contributed about $6 million. An additional $1 million of the fund will be used for future health problems linked to the Love Canal exposures.

United States v. Interlake, Inc., 429 F. Supp. 193, 432 F. Supp 987, United States District Court, Northern District of Illinois (1977)

EPA sought injunctive relief to enforce a state implementation plan. Specifically the agency aimed to prohibit Interlake from Clean Air Act violations involving its coke production facility in Chicago. The defendant successfully obtained a stay of the order while issues were being litigated in state courts. Resolution of these issues could avoid addressing constitutional issues in federal court. The defendant was simultaneously a defendant in an action brought by the Illinois EPA before the Illinois Pollution Control Board. Interlake argued that the regulation was unconstitutionally vague since a control device previously approved by the federal and state agencies was now being disapproved. Interlake hoped to get a variance or have the regulation construed narrowly so it would be found in compliance. The stay was granted despite the government's right to enforce the state plan concurrently in federal court.

The same court subsequently upheld the authority of the United States to "blacklist" noncriminal violators such as Interlake from government purchases.

United States v. Kennecott Copper Corp., 7 ER 1597, CIV-76-677 PHX-WPC (1977)

The New York corporation's Arizona-based Ray Mines Division admitted having violated an effluent discharge permit on three separate occasions by discharging leach water into Mineral Creek, a tributary of the Gila River. Although up to 10,000 gallons of leach water were discharged, prompt corrective action prevented significant damage to the environment and human health. Nonetheless, the district court fined Kennecott a total of $6,000 for the violations (the plaintiff had sought a $10,000 fine for each of the violations).

United States v. Matula and Nugent, H-82-206-1, United States District Court, Southern District of Texas (1982)

Two employees of a Mexican firm allegedly defrauded three American companies in late 1980 and early 1981 by charging the companies for supposedly shipping hazardous wastes and PCBs to Mexico for processing and proper storage. The indicted employees, Ivan Matula and Graham Nugent, allegedly arranged for storage in a Hidalgo, Texas, warehouse that did not have the authority to store such materials. Manifestos returned to the American companies wrongly indicated that the shipments had arrived in Mexico. In September, 1982, the defendants were charged in district court with violating the Resource Conservation and Recovery Act and with committing mail fraud. Defendants entered pleas of "not gulity" to all charges, and both men were acquitted in October 1983.

United States v. Oxford Royal Mushroom Products, Inc., 405 F.Supp 578 (1979)

In May, 1978, and March–April, 1979, the Pennsylvania-based mushroom producer allegedly discharged wastewater from its mushroom cannery into Big Elk Creek, a navigable water of the United States. On September 24, 1979, a federal grand jury charged both the president and general manager of Oxford Royal with negligently discharging the pollutants without having obtained the appropriate permits in violation of the Federal Water Pollution Control Act. Though the defendants argued that the government had failed to follow established proceedings in their investigation, they were ultimately convicted (in United States District Court) of the charges. On March 4, 1980, sentences for the defendants were suspended, and the defendants were placed on five years' probation. The court imposed fines totaled $100,000.

United States v. Reserve Mining Company, 294 Minn. 300 (1972), 514 F.2d 492 (1975); 417 F. Supp 789, aff'd, 543 F.2d 1210 (1976); 423 F.Supp. 759 (1976); 431 F. Supp 1248 (1977)

The Reserve Mining case was long and complex, covering seven years of conference and conciliation, traditional abatement orders, negotiations and litigation, both by and against the Reserve Mining Company located on the north shore of Lake Superior, in Silver Bay, Minnesota. The saga began with Reserve's challenge to Minnesota's compliance with a United States Interior Department requirement of preventing degradation of interstate waters. Minnesota countered with the charge that Reserve was causing a public nuisance, and the court battles began, interspersed with negotiation attempts. Reserve was charged with polluting Lake Superior by discharging taconite tailings into the lake, the primary source of drinking water for north shore residents. The specific counts included violations of the Federal Water Pollution Control Act, the Refuse Act of 1899, air pollution regulations, and the federal common law of nuisance. At issue at first was the detrimental aesthetic effects of the discharge. Later, effects on public health associated with the asbestos-like fibers composing the taconite tailings became of central importance.

The case was one which the federal court of appeals characterized as "on the frontiers of scientific knowledge." In it the court wrestled with the notion of "endangerment" of public health. On the basis of risk to public health caused by exposures to fibers in the air and water, the court mandated abatement of the disposal into the lake within a reasonable period of time. Ultimately, the case was resolved, with Reserve agreeing to terminate discharges into Lake Superior and Reserve and Minnesota agreeing to construct a $370 million facility for on-land disposal of the tailings. Reserve also paid for filtering water for north shore cities which used Lake Superior for drinking water.

United States v. Ward, 676 F.2d 94 (1982) cert. denied, in 459 US, 83-63-CIV-5 United States District Court, Eastern District of North Carolina (May 14, 1984)

The president of Ward Transformer Company was charged with unlawful disposal of toxic substances including PCBs along 200 miles of roads in rural North Carolina. Ward was convicted on eight counts of unlawful disposal of toxic substances in violation of the Toxic Substances Control Act and of aiding and abetting the unlawful disposal of toxic substances. The violations resulted from a business venture between Ward and a "long-time friend" (see *North Carolina v. Burns* above). In state court Ward had been acquitted on charges of malicious damage to real property, but the federal court charges did not require a showing of malice and damage to real property, so the conviction did not run afoul of the prohibition of double jeopardy.

Once contaminated soil had been removed from the roadside by state officials, the federal government instituted an action on January 26, 1983, against Ward and the Ward Transformer Company seeking to recover funds expended for the cleanup. Ward subsequently sought a complaint against two other firms, claiming that since the other firms were "joint-venturers" in the dumping, they should pay costs in proportion to the amount of fluid dumped. In May, 1984, the court held that a third-party action for contribution by the primary defendant is not precluded by collateral estoppal, or by CERCLA.

BIBLIOGRAPHY

Ackerman,B. A., and Hassler, W. T. *Clean Coal/Dirty Air: Or How the Clean Air Act Became a Multi-Billion-Dollar Bail-out for High-sulphur Coal Producers and What Should Be Done About it.* New Haven: Yale University Press, 1981.

Adenaes, J. The General Preventive Effects of Punishment. *University of Pennsylvania Law Review,* 1966, *114,* 949–983.

Allison, G. T. *Essence of Decision: Explaining the Cuban Missile Crisis.* Boston: Little, Brown, 1971.

Alpert, G. P. Prisons as Formal Organizations: Compliance Theory in Action. *Sociology and Social Research,* 1978, *63,* 112–130.

Amy, D. J. The Politics of Environmental Mediation. *Ecology Law Quarterly,* 1983, *11,* 1–19.

Anderson, F. R., Kneese, A. V., Reed, P. D., Taylor, S., and Stevenson, R. B. *Environmental Improvement Through Economic Incentives.* Baltimore: Johns Hopkins University Press, 1977.

Anderson, F. R., Mandelker, D. R., and Tarlock, D. A. *Environmental Protection: Law and Policy.* Boston: Little, Brown, 1984.

Anderson, J. E. Public Economic Policy and the Problems of Compliance: Notes for Research. *Houston Law Review,* 1966, *4,* 62–72.

Argyris, C., *et al. Regulating Business: The Search for an Optimum.* San Francisco: Institute for Contemporary Studies, 1978.

Arthur D. Little, Inc. *Profiles of Environmental Auditing Programs.* Cambridge: Center for Environmental Assurance, undated.

Arthur D. Little, Inc. *A Survey of Environmental Auditing.* Cambridge: Center for Environmental Assurance, undated.

Aubert, V., Ed. *Sociology of Law.* Harmondsworth, England: Penguin Books, 1969.

Bacow, L. S., and Wheeler, M. *Environmental Dispute Resolution.* New York: Plenum Press, 1984.

Bailey, G. E., and Thayer, P. S. *California's Disappearing Coast: A Legislative Challenge.* Berkeley, Cal.: Institute of Governmental Studies at University of California, 1971.

Balch, G. I. The Stick, the Carrot and Other Strategies. *Law and Policy Quarterly,* 1980, *2,* 35–59.

Ball, H. V., and Friedman, L. M. The Use of Criminal Sanctions in the Enforcement of Economic Legislation: A Sociological View. *Stanford Law Review,* 1965, *17,* 97–223.

Ball, H. V., Simpson, G. E., and Ikeda, K. Law and Social Change: Sumner Reconsidered. *American Journal of Sociology,* 1982, *67,* 532–540.

Baram, M. S. *Alternatives to Regulation: Managing Risks to Health, Safety and the Environment.* Lexington, Mass.: Lexington Books, 1982.

Bardach, E., and Kagan, R. *Going by the Book: The Problem of Regulatory Unreasonableness.* Philadelphia: Temple University Press, 1982.

Baron, B. R., and Baron, P. A Regulatory Compliance Model. *Journal of Contemporary Business*, 1980, *19*, 139–150.

Bartlett, R. V. *The Reserve Mining Controversy: Science, Technology and Environmental Quality*. Bloomington: Indiana University Press, 1980.

Bean, F. D., and Cushing, R. T. Criminal Homicide, Punishment and Deterrence: Methodological and Substantive Reconsiderations. *Social Science Quarterly*, 1971, *52*, 277–289.

Beckenstein, A. R., and Gabel, H. L. *Organizational Compliance Processes and the Efficiency of Antitrust Enforcement*. Paper presented at the annual meeting of the Law and Society Association, June, 1980.

Beckenstein, A. R., and Gabel, H. L. *A Theory of Antitrust Enforcement and Compliance*. Mimeograph, 1981.

Becker, G. S., and Stigler, G. J. Law Enforcement, Malfeasance, and Compensation of Enforcers. *The Journal of Legal Studies*, 1974, *3*, 1–18.

Bermant, G. *Implementation of Court Orders for Institutional Reform*. Paper presented at the annual meeting of the Law and Society Association, San Francisco, May, 1979.

Bernstein, M. H. *Regulating Business by Independent Commission*. Princeton: Princeton University Press, 1955.

Bilkey, W. J. Harmonizing of Economic and Environmental Needs Via a Consensus Approach: The Wisconsin Copper Mining Case. *Review of Social Economy*, 1982, *40*, 133–146.

Black, D. J. *The Behavior of Law*. New York: Academic Press, 1976.

Blau, P. M. *The Dynamics of Bureaucracy: A Study of Interpersonal Relations in Two Government Agencies*. Chicago: University of Chicago Press, 1955.

Blumrosen, A. W. Six Conditions for Meaningful Self-Regulation. *American Bar Association Journal*, 1983, *69*, 1264–1269.

Blumstein, A., Cohen, J., and Nagin, D., Eds. *Deterrence and Incapacitation: Estimating the Effects of Criminal Sanctions on Crime Rates*. Washington: National Academy of Science, 1978.

Bonine, J. E., and McGarity, T. O. *The Law of Environmental Protection*. St. Paul, Minn.: West Publishing, 1984.

Boydstun, J. E. *San Diego Field Interrogation: Final Report*. Washington, D.C.: Police Foundation, 1975.

Braithwaite, J. The Limits of Economism in Controlling Harmful Corporate Conduct. *Law and Society Review*, 1981, *16*, 481–504.

Braithwaite, J. Enforced Self-Regulation: A New Strategy for Corporate Crime Control. *Michigan Law Review*, 1982, *80*, 1466–1507.

Braithwaite, J. *Corporate Crime in the Pharmaceutical Industry*. Boston: Routledge & Kegan Paul, 1984.

Braithwaite, J. *Taking Responsibility Seriously: Corporate Compliance Systems*. Mimeograph, in press.

Braithwaite, J., and Geis, G. On Theory and Action for Corporate Crime Control. *Crime and Delinquency*, 1982, *28*, 292–314.

Braybrooke, D., and Lindblom, C. E. *A Strategy of Decision: Policy Evaluation as a Social Process*. New York: Free Press of Glencoe, 1963.

Breyer, S. Analyzing Regulatory Failure: Mismatches, Less Restrictive Alternatives and Reform. *Harvard Law Review*, 1979, *92*, 549–609.

Brigham, J., and Brown, D. W. Distinguishing Penalties and Incentives. *Law and Policy Quarterly*, 1980, *2*, 5–9.

Brooks, H. Knowledge and Action—Dilemma of Science Policy in the 70's, *Daedalus*, 1973, p. 125.

Brooks, H. Expertise and Politics—Problems and Tensions. *Proceedings of the American Philosophical Society*, 1975, *119*, 257.

Brooks, H. The Resolution of Technically Intensive Public Policy Disputes. *Science, Technology, and Human Values*, 1984a, *46*, 39–50.

Brooks, H. *Public-Private Partnership: New Opportunities for Meeting Social Needs*. Cambridge, Mass.: Ballinger, 1984b.

Brown, D. W., and Stover, R. V. Court Directives and Compliance. *American Politics Quarterly*, 1977, *5*, 465–480.

Bruff, H. H. Presidential Power and Administrative Rulemaking. *Yale Law Journal*, 1979, *88*, 451–508.

Bryan, J. T. The Economic Inefficiency of Corporate Criminal Liability. *Journal of Criminal Law and Criminology*, 1982, *73*, 582–603.

Bryden, D. P. Environmental Rights in Theory and Practice. *Minnesota Law Review*, 1976, *62*, 163–228.

Cardin, M., and Brilliant, L. B. The Search for Effective State Decision-Making About Toxic Substances: Michigan's Toxic Substances Control Commission Act. *Wayne Law Review*, 1979, *25*, 1217–1249.

Chambliss, W. J. The Deterrent Influence of Punishment. *Crime and Delinquency*, 1966, *12*, 70–75.

Champagne, A., and Nagel, S. Law and Social Change. In E. Seidman, Ed., *Handbook of Social Intervention*. Beverly Hills, Cal.: Sage, 1983.

Child, J. Strategies of Control and Organizational Behavior. *Administrative Science Quarterly*, 1973, *18*, 1–17.

Christiansen, G., Gollop, F., and Haverman, R. *Environmental and Health/Safety Regulations, Productivity, Growth, and Economic Performance*. Washington, D.C.: Government Printing Office, 1980.

Clay, T. *Combating Cancer in the Workplace: Implementation of the California Occupational Carcinogens Control Act*. Ph.D. Dissertation, University of California, Irvine, 1983.

Clinard, M. B. *Corporate Ethics and Crime: The Role of Middle Management*. Beverly Hills, Cal.: Sage, 1983.

Clinard, M. B., and Yeager, P. C. *Corporate Crime*. New York: Free Press, 1980.

Coffee, J. C. No Soul to Damn. No Body to Kick: An Unscandalized Inquiry into the Problem of Corporate Punishment. *Michigan Law Review*, 1981, *79*, 386–459.

Comment. Substantial Compliance: Judicial Amendment to NEPA, *EDF v. Andrus*. *University of Colorado Law Review*, 1982, *53*, 347–366.

Committee on Environmental Controls, Section of Corporation Banking and Business Law, American Bar Association. Structuring Corporate Compliance Programs for Pollution Control Requirements. *Business Lawyer*, 1980, *35*, 1459–1491.

Conybeare, J. A. Politics and Regulation: The Public Choice Approach. *Australian Journal of Public Administration*, 1982, *41*, 33–45.

Coombs, F. S. The Bases of Noncompliance with a Policy. *Policy Studies Journal*, 1980, *8*, 885–892.

Costle, D. M. Environmental Regulation and Regulatory Reforms. *Washington Law Review*, 1982, *57*, 409–432.

Council on Environmental Quality. *Environmental Quality—1980*. Washington, D.C.: Government Printing Office, 1980.

Crandall, R. W. *Controlling Industrial Pollution: The Economics and Politics of Clean Air*. Washington, D.C.: Brookings Institution, 1983.

Crandall, R. W., and Lave, L. B., Eds. *The Scientific Basis of Health and Safety Regulation*. Washington, D.C.: Brookings Institution, 1981.

Currie, D. P. Air Pollution Control in West Germany. *University of Chicago Law Review*, 1982, *49*, 355–393.

Curzan, M. P., and Pelesh, M. L. The Changing Role of Outside Counsel: A Proposal for a Legal Audit. *The Notre Dame Lawyer*, 1981, *56*, 838–849.

Cutler, L. N., and Johnson, D. R. Regulation and the Political Process. *Yale Law Journal*, 1975, *84*, 1395–1418.

Cyert, R. M., and March, J. G. *A Behavioral Theory of the Firm*. Englewood Cliffs, N.J.: Prentice-Hall, 1963.

Daneke, G. A. The Future of Environmental Protection: Reflections on the Difference Between Planning and Regulating. *Public Administration Review*, 1982, *42*, 227–233.

Darley, J. M., and Latane, B. Bystander Intervention in Emergencies: Diffusion of Responsibility. *Journal of Personality and Social Psychology*, 1968, *8*, 377.

DeLong, J. V. Informal Rulemaking and the Integration of Law and Policy. *Virginia Law Review*, 1982, *65*, 257–356.

DiMento, J. F. *Managing Environmental Change: A Legal and Behavioral Perspective*. New York: Praeger, 1976.

DiMento, J. F. Citizen Environmental Litigation and the Administrative Process: Empirical Findings, Remaining Issues and a Direction for Future Research. *Duke Law Journal*, 1977, *1977*, 409–448.

DiMento, J. F. *The Consistency Doctrine and the Limits of Planning*. Cambridge: Oelgeschlager, Gunn & Hain, 1980.

DiMento, J. F., Dozier, M. D., Emmons, S. L., Hagman, D. G., Kim, C., Greenfield-Sanders, K., Waldau, P. F., and Woollacott, J. A. Land Development and Environmental Control in the California Supreme Court: The Deferential, the Preservationist and the Preservationist-Erratic Eras. *UCLA Law Review*, 1980, *27*, 859–1066.

DiMento, J., Lambert, W., Suarez-Villa, L., and Tripodes, J. Siting Low Level Nuclear Waste Facilities. *Journal of Environmental Systems* 1985–86, *15*(1), 19–43.

Dinkens, C. E. Shall We Fight or Shall We Finish: Environmental Dispute Resolution in a Litigious Society. *Environmental Law Reporter*, 1984, *14*, 10398–10401.

Diver, C. S. The Assessment and Mitigation of Civil Money Penalties by Federal Administrative Agencies. *Columbia Law Review*, 1979, *79*, 1435–1502.

Diver, C. S. A Theory of Regulatory Enforcement. *Public Policy*, 1980, *28*, 257–299.

Diver, C. S. The Optimal Precision of Administrative Rules. *Yale Law Journal*, 1983, *93*, 65–109.

Dolbeare, K. M., and Hammond, P. E. *The School Prayer Decisions from Court Policy to Local Practice*. Chicago: Chicago University Press, 1971.

Dolgin, E. L., and Gilbert, T., Eds. *Federal Environmental Law*. St. Paul: West Publishing, 1974.

Doolittle, F. C. *Land-use Planning and Regulation on the California Coast: The State Role*. Davis, Cal.: University of California, Environmental Quality Series, 1972.

Downing, P. B., and Watson, W. D. The Economics of Enforcing Air Pollution Controls. *Journal of Environmental Economics and Management*, 1974, *1*, 219–250.

Drayton, W. Economic Law Enforcement. *Harvard Environmental Law Review*, 1980, *4*, 1–40.

Dunlop, J. T. The Limits of Legal Compulsion. *Labor Law Journal*, 1976, *27*, 67–74.

Edelman, L. H., and Walline, R. E. Developing a Cooperative Approach to Environmental Regulation. *Natural Resources Lawyer*, 1984, *16*, 489–497.

Efron, E. *The Apocalyptics: Cancer and the Big Lie*. New York: Simon & Schuster, 1984.

Emery, F. E., and Trist, E. L. The Causal Texture of Organizational Environments. *Human Relations*, 1965, *18*, 21–32.

Engelberg, D. *Controlling Workplace Carcinogens: The Impact of Evidentiary Uncertainty Upon Regulatory Effectiveness*. Doctoral dissertation, Michigan State University, 1982.

Environmental Protection Agency. *Health Consequences of Sulfur Dioxide: A Report from CHESS, 1970–1971*. EPA-650/1-74-004. May, 1974.

Environmental Protection Agency. *Private Sector Provision of Operation and Maintenance Services to Publicly Owned Treatment Plants*. Urban Systems Research. July, 1980.

Erickson, M. L., and Gibbs, J. P. Specific Versus General Properties of Legal Punishments and Deterrence. *Social Science Quarterly,* 1975, *56,* 390–397.

Erlich, I. Participation in Illegitimate Activities: A Theoretical and Empirical Investigation. *Journal of Political Economy,* 1973, *81,* 521–565.

Ermann, M. D., and Lundman, R. J. Deviant Acts by Complex Organizations; Deviance and Social Control at the Organizational Level of Analysis. *Sociological Quarterly,* 1975, *19,* 55–67.

Evan, W. M. Law as an Instrument of Social Change. In A. M. Gouldner and S. M. Miller, Eds., *Applied Sociology: Opportunities and Problems.* New York: Free Press, 1965.

Faley, T., and Tedeschi, J. T. Status and Reactions to Threats. *Journal of Personality and Social Psychology,* 1971, *17,* 192.

Feeley, M. Coercion and Compliance: A New Look at an Old Problem. In S. Krislov, K. O. Boyum, H. N. Clark, R. C. Schafer, and S. O. White, Eds., *Compliance and the Law: A Multidisciplinary Approach.* Beverly Hills, Cal.: Sage, 1972.

Festinger, L. *A Theory of Cognitive Dissonance.* Stanford: Stanford University Press, 1964.

Findley, R. W., and Farber, D. A. *Environmental Law: Cases and Materials.* St. Paul: West Publishing, 1981.

Fiorina, M. P. Legislative Choice of Regulatory Forms: Legal Process or Administrative Process? *Public Choice,* 1982, *39,* 33–66.

Fisse, W. B. The Use of Publicity as a Criminal Sanction Against Business Corporations. *Melbourne University Law Review,* 1971, *8,* 107–150.

Fisse, W. B. The Social Policy of Corporate Criminal Responsibility. *Adelaide Law Review,* 1978, *6,* 361–412.

Fisse, W. B. Community Service as a Sanction Against Corporations. *Wisconsin Law Review,* 1981, *1981,* 970–1017.

Fisse, W. B. Reconstructing Corporate Criminal Law: Deterrence, Retribution, Fault and Sanctions. *Southern California Law Review,* 1982, *56,* 1141–1246.

Fisse, W. B., and Braithwaite, J. *The Impact of Publicity on Corporate Offenders.* Albany: State University of New York Press, 1983.

Freeman, L. Innovations in Compliance. *Civil Rights Digest,* 1978, *16,* 21–27.

Freilich, R. H. Solving the ''Taking'' Equation: Making the Whole Equal the Sum of the Parts. *Institute on Planning, Zoning and Eminent Domain,* 1982, pp. 301–353.

Frieden, B. J. *The Environmental Protection Hustle.* Cambridge: MIT Press, 1979.

Friedman, L. M. *The Legal System: A Social Science Perspective.* New York: Russell Sage Foundation, 1975.

Garner, L. Management Control in Regulatory Agencies: A Modest Proposal for Reform. *Administrative Law Review,* 1982, *34,* 465–481.

Gatti, J., Ed. *The Limits of Government Regulation.* New York: Academic Press, 1981.

Geis, G. Deterring Corporate Crime. In M. D. Ermann and R. J. Lundman, Eds., *Corporate and Governmental Deviance: Problems of Organizational Behavior in Contemporary Society.* New York: Oxford University Press, 1978.

Geis, G., and Meier, R., Eds. *White-Collar Crime: Offenses in Business, Politics and the Professions.* New York: Free Press, 1977.

Getman, J. G. *The Relationship Between Law and Private Ordering in Labor Relations.* Paper presented at the annual meeting of the Law and Society Association, San Francisco, May 11, 1979.

Gibbs, J. P. Crime, Punishment and Deterrence. *Southwestern Social Science Quarterly,* 1968, *48,* 515–530.

Ginsburg, D. H., and Abernathy, W. J. *Government, Technology and the Future of the Automobile.* New York: McGraw-Hill, 1978.

Glenn, M. K. The Crime of "Pollution": The Role of Federal Water Pollution Criminal Sanctions. *The American Criminal Law Review*, 1973, *11*, 835–882.

Goldschmid, H. J. An Evaluation of the Present and Potential Use of Civil Money Penalties as a Sanction by Federal Administrative Agencies. *Recommendations and Reports of the Administrative Conference of the United States*, 1972, *2*, 896–964.

Goldstein, B. H., and Howard, H. H. Antitrust Law and the Control of Auto Pollution: Rethinking the Alliance Between Competition and Technical Progress. *Environmental Law*, 1980, *10*, 517–558.

Gray, L. N., and Martin, J. D. Punishment and Deterrence: Another Analysis of Gibbs' Data. *Social Science Quarterly*, 1969, *50*, 389–395.

Greenwald, A. G. Promotive Law-Compliance Scheduling: New Perspectives for Environmental Law. *Natural Resources Lawyer*, 1977, *10*, 729–736.

Greer, C. R., and Downey, H. K. Industrial Compliance with Social Legislation: Investigations of Decision Rationals. *Academy of Management Review*, 1982, *7*, 488–498.

Grenier, G. On Compliance With the Minimum-Wage Law. *Journal of Political Economy*, 1982, *90*, 184–187.

Grobstein, C. Utilization of Science in Formulating Legislative and Regulatory Policy. *Equilibrium*, 1983, *11*, 2–7.

Groening, W. A. *The Modern Corporate Manager: Responsibility and Regulation*. New York: McGraw-Hill, 1981.

Gross, E. Organizational Crime: A Theoretical Perspective. *Studies in Symbolic Interaction*, 1978, *1*, 55–85.

Grossman, J. B., and Grossman, M. H., Eds. *Law and Change In Modern America*. Pacific Palisades: Goodyear Publishing, 1971.

Gruhl, J. The Supreme Court's Impact on the Law of Libel: Compliance by Lower Federal Courts. *Western Political Quarterly*, 1981, *33*, 502–519.

Gunningham, N. *Pollution, Social Interest and the Law*. London: M. Robertson, 1974.

Hall, R. M. The Evolution and Implementation of EPA's Regulatory Program to Control the Discharge of Toxic Pollutants to the Nation's Waters. *Natural Resources Lawyer*, 1977, *10*, 507–529.

Hanf, K. Regulatory Structures: Enforcement as Implementation. *European Journal of Political Research*, 1982, *10*, 159–172.

Harter, P. J. Negotiating Regulations: A Cure for Malaise. *Georgetown Law Journal*, 1982, *71*, 1–118.

Harter, P. J. The Political Legitimacy and Judicial Review of Consensual Rules. *American University Law Review*, 1983, *32*, 471–496.

Hawkins, K. *Enforcing Water Pollution Law in England*. Oxford: Mimeograph, 1980.

Hawkins, K. Bargain and Bluff: Compliance Strategy and Deterrence in the Environment of Regulation. *Law and Policy Quarterly*, 1983, *5*, 35–73.

Hawkins, K. *Environment and Enforcement: Regulation and the Social Definition of Pollution*. New York: Oxford University Press, 1984.

Haynes, J. K. Michigan's Environmental Protection Act in Its Sixth Year: Substantive Environmental Law from Citizens' Suits. *Journal of Urban Law*, 1976, *53*, 590–700.

Hearings Before the Committee on Environment and Public Works, United States Senate, 98th Congress, First Session. February 15, 1983. Washington, D.C.: Government Printing Office.

Heller, K., and Monahan, J. *Psychology and Community Change*. Homewood, Ill.: Dorsey Press, 1977.

Henderson, J. A., and Pearson, R. N. Implementing Federal Environmental Policies: The Limits of Aspirational Commands. *Columbia Law Review*, 1978, *78*, 1429–1470.

Hileman, B. Environmental Dispute Resolution. *Environmental Science and Technology*, 1983, *17*, 165A–168A.

Hohenemser, C., Kates, R. W., and Slovic, P. The Nature of Technological Hazard. *Science*, 1983, *220*, 378–384.

Horai, J., and Tedeschi, J. T. Effects of Credibility and Magnitude of Punishment on Compliance to Threats. *Journal of Personality and Social Psychology*, 1969, *12*, 164–169.

Huber, P. The Market for Risk. *Regulation*, March, 1984, 33–40.

Ippolito, R. A. The Effects of Price Regulation in the Automobile Insurance Industry. *Journal of Law and Economics*, 1979, *12*, 55–89.

Jacob, H. Deterrent Effects of Formal and Informal Sanctions. *Law and Policy Quarterly*, 1980, *2*, 61–80.

Johnson, M. B. Introduction. In H. M. Brown, Ed., *The California Coastal Plan: A Critique*. San Francisco: Institute for Contemporary Studies, 1976.

Johnson, R. M. *The Dynamics of Compliance: Supreme Court Decision-Making from a New Perspective*. Evanston, Ill.: Northwestern University Press, 1967.

Jones, C. O. *An Introduction to the Study of Public Policy*. North Scituate, Mass.: Duxbury Press, 1977.

Jones, H. W. *The Efficacy of Law*. Evanston, Ill.: Northwestern University Press, 1969.

Kadish, S. H. Some Observations on the Use of Criminal Sanctions in Enforcing Economic Regulations. *University of Chicago Law Review*, 1963, *30*, 423–449.

Kagan, R. *Regulatory Enforcement Styles*. Paper presented at the annual meeting of the Law and Society Association. Boston, June 7–10, 1984.

Kagan, R., and Scholz, R. The Criminology of the Corporation and Regulatory Enforcement Strategies. *Jahrbuch fhur Rechtssoziologie und Rechtstheorie*, 1981, *7*, 352–377.

Kantrowitz, A. Proposal for an Institution for Scientific Judgement. *Science*, 1967, *156*, 763–767.

Kaplan, S. A. Some Ruminations of the Role of Counsel for a Corporation. *Notre Dame Lawyer*, 1981, *56*, 873–886.

Kapp, R. W. Some Problems of a Legal Compliance Audit. *Business Lawyer*, 1978, *56*, 2467–2479.

Katz, D., and Kahn, R. L. *The Social Psychology of Organizations* (2nd ed.). New York: Wiley, 1978.

Kelling, G. L., et al. *The Kansas City Preventive Patrol Experiment*. Washington, D.C.: Police Foundation, 1974.

Kelman, S. Occupational Safety and Health Administration. In J. Q. Wilson, Ed., *The Politics of Regulation*. New York: Basic Books, 1980.

Kolko, G. *Railroads and Regulation 1877–1916*. Princeton: Princeton University Press, 1965.

Kovel, A. A Case for Civil Penalties: Air Pollution Control. *Journal of Urban Law*, 1969, *46*, 153–171.

Krier, J. E. Air Pollution and Legal Institutions: An Overview. In P. B. Downing, Ed., *Air Pollution and the Social Sciences: Formulating and Implementing Control Programs*. New York: Praeger, 1971.

Krislov, S. The Parameters of Power: The Concept of Compliance as an Approach to the Study of the Legal and Political Process. In S. Krislov, K. O. Boyum, H. N. Clark, R. C. Schaefer, and S. O. White, Eds., *Compliance and the Law: A Multidisciplinary Approach*. Beverly Hills, Cal. Sage, 1972.

Lake, L. M. The Environmental Mandate: Activities and the Electorate. *Political Science Quarterly*, 1983, *2*, 215–233.

Landis, J. M. *Report on Regulatory Agencies to the President-Elect*. Washington, D.C.: Government Printing Office, 1960.

Lane, R. E. *The Regulation of Businessmen—Social Conditions of Government Economic Control.* Hamden: Anchor Books, 1954.

LaPorte, T. R. *Organizational Response to Complexity: Research and Development as Organized Inquiry and Action: Part I.* Working Paper No. 141. Berkeley, Cal.: Center for Planning and Development, Institute of Urban and Regional Development, 1971.

Latin, H. A. The Feasibility of Occupational Health Standards: An Essay on Legal Decisionmaking Under Uncertainty. *Northwestern University Law Review,* 1983, *75,* 585–631.

Lave, L. B. *The Strategy of Social Regulation: Decision Frameworks for Policy.* Washington, D.C.: Brookings Institution, 1981.

Lave, L. B., and Omenn, G. S. *Cleaning the Air: Reforming the Clean Air Act.* Washington, D.C.: Brookings Institution, 1981.

Lawrence, R. R., and Lorsch, J. W. *Organization and Environment: Managing Differentiation and Integration.* Homewood, Ill.: R. D. Irwin, 1969.

Leonard, H. J. Environmental Regulations, Multinational Corporations and Industrial Development in the 1980's. *Habitat International* 1982, *6,* 323–341.

Leone, R. A. The Real Costs of Regulation. *Harvard Business Review,* 1977, *55,* 57–66.

Levin, M. H. Crimes Against Employees: Substantial Criminal Sanctions Under the Occupational Safety and Health Act. *The American Criminal Law Review,* 1977, *14,* 717–745.

Levine, A. G. *Love Canal: Science, Politics and People.* Lexington: Lexington Books, 1982.

Levine, J. P. Methodological Concerns in Studying Supreme Court Efficacy. *Law and Society Review,* 1970, *4,* 583–611.

Likens, T. W., and Kohfeld, C. W. *Models of Mass Compliance: Contextual or Economic Approach.* Mimeograph, undated.

Lindblom, C. E. The Science of "Muddling Through." *Public Administration Review,* 1959, *19,* 77–88.

Lippitt, R., Watson, J., and Westley, B. *The Dynamics of Planned Change: A Comparative Study of Principles and Techniques.* New York: Harcourt, Brace, 1958.

Liroff, R. NEPA-Where Have We Been and Where Are We Going? *Journal of American Planning Association,* 1980, *47,* 154–157.

Lowi, T. J. *The End of Liberalism: Ideology, Principles and the Crisis of Public Authority.* New York: Norton, 1969.

Lund, L. *Corporate Organization for Environmental Policymaking.* New York: Conference Board, 1977.

Luschbough, H. A Food Company's Approach to the Inspectional Process. *Food, Drug and Cosmetic Law Journal,* 1980, *35,* 436–450.

MacAvoy, P. W. *The Economic Effects of Regulation: The Trunk-Line Railroad Cartels and the Interstate Commerce Commission Before 1900.* Cambridge: M.I.T. Press, 1965.

Magat, W. A. The Effects of Environmental Regulation on Innovation. *Law and Contemporary Problems,* 1979, *43,* 4–25.

Magat, W. A. *Reform of Environmental Regulation.* Cambridge: Ballinger, 1982.

Majone, G. Process and Outcome in Regulatory Decision-Making. *American Behavioral Scientist,* 1979, *22,* 561–583.

Malone, M. T., and McCormick, R. E. A Positive Theory of Environmental Quality Regulation. *Journal of Law and Economics,* 1982, *25,* 99–123.

Mann, K., Wheeler, S., and Sarat, A. Sentencing the White-Collar Offender. *American Criminal Law Review,* 1980, *17,* 479–500.

Marcus, A. Environmental Protection Agency. In J. Q. Wilson, Ed., *The Politics of Regulation.* New York: Basic Books, 1980.

Marin, A. Pollution Control: Economists' Views. *The Three Banks Review,* 1979, *121,* 21–41.

Marino, K. E. *An Investigation of the Relationships Between Affirmative Action Compliance Charac-*

teristics of Organizational Structure, and Managerial Values/Attitudes. Doctoral dissertation, University of Massachusetts, 1978.

Marlow, M. L. The Economics of Enforcement: The Case of OSHA. *Journal of Economics and Business,* 1982, *34,* 164–171.

Matheny, A. R., and Williams, B. A. Scientific Disputes and Adversary Procedures in Policy Making. *Law and Policy Quarterly.* 1981, *3,* 341–364.

Maupin, M. W. Environmental Law, the Corporate Lawyer and the Model Rules of Professional Conduct. *Business Lawyer,* 1981, *36,* 431–460.

Mayda, J. The Penal Protection of the Environment. *American Journal of Corporate Law,* 1978, *26,* 471–480.

Mazmanian, D. A., and Sabatier, P., Eds. *Effective Policy Implementation.* Lexington: Lexington Books, 1981.

McAllister, A. M. Evaluation in Environmental Planning: *Assessing Environmental, Social, Economic and Political Tradeoffs.* Cambridge: M.I.T. Press, 1980.

McCaffrey, D. P. Corporate Resources and Regulatory Pressures: Toward Explaining a Discrepancy. *Administrative Science Quarterly,* 1982, *27,* 398–419.

McCraw, T. K. Regulation in America: A Review Article. *Business History Review,* 1975, *49,* 159–183.

McDermott, M. F. Occupational Disqualification of Corporate Executives: An Innovative Condition of Probation. *Journal of Criminal Law and Criminology,* 1982, *73,* 604–641.

McGarity, T. A. Substantive and Procedural Discretion in Administrative Resolution of Science Policy Questions: Regulating Carcinogens in EPA and OSHA. *Georgetown Law Journal,* 1979, *67,* 729–810.

McGlennon, J. Negotiating Regulation: Success at EPA. In ALI–ABA Course of Study Materials. *Environmental Law.* Philadelphia: American Law Institute, 1985.

McKean, R. N. Enforcement Costs in Environmental and Safety Regulations. *Policy Analysis,* 1980, *6,* 269–289.

McKie, J. W., Ed. *Social Responsibility and the Business Predicament.* Washington, D.C.: Brookings Institution, 1974.

McNolds, D., Unkovic, J. C., and Levin, J. L. Thoughts on the Role of Penalties in the Enforcement of the Clean Air and Clean Water Acts. *Duquesne Law Review,* 1978, *17,* 1–31.

Meier, K. J., and Morgan, D. R. Citizen Compliance with Public Policy: The National Maximum Speed Law. *Western Political Quarterly,* 1982, *35,* 258–273.

Meier, R. F., and Johnson, W. T. Deterrence as a Social Control: The Legal and Extralegal Production of Conformity. *American Sociological Review,* 1977, *42,* 292–304.

Melnick, R. S. Deadlines, Common Sense and Cynicism. *The Brookings Review,* 1983a, *2,* 21–24.

Melnick, R. S. *Regulation and the Courts: The Case of the Clean Air Act.* Washington, D.C.: Brookings Institution, 1983b.

Mendelsohn, [n.i.]. *Regulation Through Prices or Quantities.* Mimeograph, December, 1978.

Michael, D. N. *On Learning to Plan and Planning to Learn.* San Francisco: Jossey-Bass, 1973.

Mileti, D. S., Gillespie, D. F., and Eitzen, D. S. Structure and Decisionmaking in Corporate Organizations. *Sociology and Social Research,* 1979, *63,* 724–744.

Milner, N. A. *The Court and Local Law Enforcement: The Impact of Miranda.* Beverly Hills, Cal.: Sage, 1971.

Milstein, I. M., and Katsh, S. M. *The Limits of Corporate Power: Existing Constraints on the Exercise of Corporate Discretion.* New York: MacMillan, 1981.

Mitnick, B. M. *The Political Economy of Regulation: Creating, Designing, and Removing Regulatory Forms.* New York: Columbia University Press, 1980.

Mitnick, B. M. The Strategic Uses of Regulation and Deregulation. *Business Horizons,* 1981, *24,* 71–83.

Mix, D. D. The Misdemeanor Approach to Pollution Control. *Arizona Law Review*, 1968, *10*, 90–96.

Moe, T. M. Regulatory Performance and Presidential Administration. *American Journal of Political Science*, 1982, *26*, 197–224.

Morris, J. S. Environmental Problems and the Use of Criminal Sanctions. *Land and Water Law Review*, 1972, *7*, 421–431.

Muir, W. K. *Prayer in the Public Schools: Law and Attitude Change*. Chicago: University of Chicago Press, 1967.

Nagel, S. S. Incentives for Compliance with Environmental Law. *American Behavioral Scientist*, 1974, *17*, 690–710.

Nagy, L., and Olson, B. Mercury in Aquatic Environments: A General Review. *Water*, 1982, *9*, 12–16.

National Research Council. *Defense and Incapacitation: Estimating the Effects of Criminal Sanctions on Crime Roles*. Washington: National Academy of Science, 1978.

Needleman, M. L. Organization Crime: Two Models of Criminogenesis. *Sociological Quarterly*, 1980, *20*, 517–528.

Neiman, M. The Virtues of Heavy-Handedness in Government. *Law and Policy Quarterly*, 1980, *2*, 11–33.

Nisbet, R. Cancer and Ideology: A Book Review. *Regulation*, 1984, March/April, 41–44.

Noll, R. G. *Breaking Out of the Regulatory Dilemma: Alternatives to the Sterile Choice*. California Institute of Technology Social Science Working Paper No. 108. January, 1976.

Note. Decisionmaking Models and the Control of Corporate Crime. *Yale Law Journal*, 1976, *85*, 1091–1129.

Note. Criminal Prosecution under the Federal Water Pollution Control Act. *Chicago-Kent Law Review*, 1980, *56*, 983–1010.

Note. The Implications of Upjohn. *Notre Dame Lawyer*, 1981, *56*, 887–902.

Note. Rethinking Regulation: Negotiation as an Alternative to Traditional Rulemaking. *Harvard Law Review*, 1981, *94*, 1871–1891.

Note. The Permissible Scope of OSHA Complaint Inspections. *University of Chicago Law Review*, 1982, *49*, 203–234.

O'Brien, T. The Use of Civil Penalties in Enforcing the Clean Water Act Amendments of 1977. *University of San Francisco Law Review*, 1978, *12*, 437–463.

Orloff, N. *Developing a Corporate Environmental Strategy*. Mimeograph, April, 1977.

Owen, B. M., and Braeutigam, R. *The Regulation Game: Strategic Use of the Administrative Process*. Cambridge: Ballinger, 1978.

Packer, H. L. *The Limits of the Criminal Sanction*. Stanford: Stanford University Press, 1968.

Paglin, M. D., and Shor, E. Regulatory Agency Responses to the Development of Public Participation. *Public Administration Review*, 1977, *37*, 140–148.

Palmer, J., and Barlett, R. Understanding Compliance and Noncompliance with Law: The Contributions of Utility Theory—A Comment. *Social Science Quarterly*, 1977, *58*, 332–335.

Parnell, A. H. Manufacturers of Toxic Substances: Tort Liability and Punitive Damages. *Forum*, 1982, *17*, 947–968.

Paternoster, R., Saltzman, L. E., Waldo, G. P., and Chiricos, T. G. Perceived Risk and Social Control: Do Sanctions Really Deter? *Law and Society Review*, 1983, *17*, 457–479.

Patton, L. K. Settling Environmental Disputes: The Experience With and Future of Environmental Mediation. *Environmental Law*, 1984, *14*, 547–554.

Pearson, J. S. Organizational Response to Occupational Injury and Disease: The Case of the Uranium Industry. *Social Forces*, 1978, *57*, 23–41.

Pennings, H. *Environment, Structure, Independence and Their Relevance for Organizational Effectiveness*. Doctoral dissertation, University of Michigan, 1973.

Pertshak, M. *Revolt Against Regulation: The Rise and Pause of the Consumer Movement.* Berkeley, Cal.: University of California Press, 1982.

Peskin, H., Portney, P. R., and Kneese, A. V. Eds. *Environmental Regulation and the U.S. Economy.* Baltimore: Johns Hopkins University Press, 1981.

Pfeffer, J., Salancik, G. R., and Leblebici, H. The Effects of Uncertainty on the Use of Social Influence in Organizational Decisionmaking. *Administrative Science Quarterly,* 1976, *21,* 227–245.

Poole, R. W. *Instead of Regulation: Alternatives to Federal Regulatory Agencies.* Lexington: Lexington Books, 1982.

Posner, R. A. The Behavior of Administrative Agencies. *Journal of Legal Studies,* 1972, *1,* 305–347.

Posner, R. A. *Economic Analysis of Law.* Boston: Little, Brown, 1973.

Posner, R. A. Theories of Economic Regulation. *The Bell Journal of Economics,* 1974, *5,* 335–358.

Pound, R. The Limits of Effective Legal Action. *Pennsylvania Bar Association Reports,* 1969, *11,* 221–239.

Preston, N. S. A Right of Rebuttal in Informal Rulemaking: May Courts Impose Procedures to Ensure Rebuttal of Ex Parte Communications and Information Derived from Agency Files After *Vermont Yankee. Administrative Law Review,* 1980, *32,* 621–665.

Quirk, P. J. *Industry Influence in Federal Regulatory Agencies.* Princeton: Princeton University Press, 1981.

Rabin, R. L., Ed. *Perspectives on the Administrative Process.* Boston: Little, Brown, 1979.

Reed, P. D. Environmental Audits and Confidentiality: Can What You Know Hurt You As Much As What You Don't Know? *Environmental Law Reporter,* 1983, *13,* 10303–10308.

Reich, R. B. Regulation by Confrontation or Negotiation ? *Harvard Business Review,* 1981, *59,* 82–93.

Reiss, A. J. Organizational Deviance. In M. D. Ermann and R. J. Lundman, Eds., *Corporate and Governmental Deviance: Problems of Organizational Behavior in Contemporary Society.* New York: Oxford University Press, 1978.

Renfrew, C. B. The Paper Label Sentences—An Evaluation. *Yale Law Journal,* 1977, *86,* 590–618.

Report Prepared for the Subcommittee on Special Studies, Investigation and Oversight, and the Subcommittee on the Environment and the Atmosphere of the Committee on Science and Technology, U.S. House of Representatives, Ninety-Fourth Congress. *The Environmental Protection Agency's Research Program With Primary Emphasis on the Community Health and Environmental Surveillance System (CHESS): An Investigative Report.* Washington, D.C.: Government Printing Office, 1976.

Richardson, G., Ogus, A., and Burrows, P. *Policing Pollution: A Study of Regulation and Enforcement.* New York: Oxford University Press, 1982.

Roberts, M. J., and Bluhm, J. S. *The Choices of Power: Utilities Face the Environmental Challenge.* Cambridge: Harvard University Press, 1981.

Roberts, J. J., Thomas, S. R., and Dowling, M. J. Mapping Scientific Disputes that Affect Public Policymaking. *Science, Technology and Human Values,* 1984, *9,* 112–122.

Robinson, G. O. On Reorganizing the Independent Regulatory Agencies. *Virginia Law Review,* 1971, *57,* 947–995.

Rodgers, H. R. *Community Conflict, Public Opinion and the Law: The Amish Dispute in Iowa.* Columbus: C. E. Merrill, 1969.

Rodgers, H. R. Law as an Instrument of Public Policy. *American Journal of Political Science,* 1973, *17,* 638–647.

Rodgers, W. H. *Handbook of Environmental Law.* St. Paul: West Publishing, 1977.

Rodgers, W. H. Benefits, Costs, and Risks: Oversight of Health and Environmental Decisionmaking. *Harvard Environmental Law Review,* 1980, *4,* 191–226.

Roiter, E. D. Illegal Corporate Practices and the Disclosure Requirements of the Federal Securities Laws. *Fordham Law Review*, 1982, *50*, 781–813.

Romano, R. Metapolitics and Corporate Law Reform. *Stanford Law Review*, 1984, *36*, 923–1016.

Rose-Ackerman, S. Effluent Charges: A Critique. *Canadian Journal of Economics*, 1973, *6*, 512–528.

Rothenberg, J. The Impact of Regulation on an Exhaustive Resource Industry: A Methodological Approach and a Model. *Law and Contemporary Problems*, 1979, *43*, 112–149.

Royston, M. G. Making Pollution Prevention Pay. *Harvard Business Review*, 1980, *58*, 6–22.

Russell, M. Regulatory Negotiation Outline. In ALI–ABA Course of Study Materials. *Environmental Law*. Philadelphia: American Law Institute, 1985.

Ryan, D. R. Environmental Regulation: A New Approach. *Environmental Management*, 1982, *16*, 95–100.

Sabatier, P. A. Social Movements and Regulatory Agencies: Toward a More Adequate and Less Pessimistic Theory of Clientele Capture. *Policy Science*, 1975, *6*, 301–342.

Sabatier, P. A. Regulatory Policy-Making: Toward a Framework of Analysis. *National Resources Journal*, 1977, *17*, 415–460.

Sabatier, P. A., and Mazmanian, D. A. *The Conditions of Effective Implementation: A Guide to Accomplishing Policy Objectives*. Mimeograph, 1978.

Sabatier, P. A., and Mazmanian, D. A. *Can Regulation Work? The Implementation of the 1972 California Coastal Initiative*. New York: Plenum Press, 1984.

Sax, J. L. *Defending the Environment: A Strategy for Citizen Action*. New York: Knopf, 1971.

Sax, J. L., and Conner, R. L. Michigan's Environmental Protection Act of 1970: A Progress Report. *Michigan Law Review*, 1972, *70*, 1004–1106.

Sax, J. L., and DiMento, J. F. Environmental Citizen Suits: Three Years' Experience Under the Michigan Environmental Protection Act. *Ecology Law Quarterly*, 1974, *4*, 1–62.

Schelling, T. C. Command and Control. In J. W. McKie, Ed., *Social Responsibility and the Business Predicament*. Washington, D.C.: Brookings Institution, 1974.

Schelling, T. C., Ed. *Incentives for Environmental Protection*. Cambridge: M.I.T. Press, 1983.

Schnapp, J. B. *Corporate Strategies of the Automotive Manufacturers*. Lexington: Lexington Books, 1979.

Schneider, K. The Data Gap: What We Don't Know About Chemicals. *The Amicus Journal*, 1985, *6*, 15–24.

Schneider, M. W. Criminal Enforcement of Federal Water Pollution Control Laws in an Era of Deregulation. *Journal of Criminal Law and Criminology*, 1982, *73*, 642–674.

Schoenbaum, T. J. *Environmental Policy Law: Cases, Readings and Text*. Mineola: Foundation Press, 1982.

Scholz, J. T. Voluntary Compliance and Regulatory Enforcement. *Law and Policy*, 1984, *6*(4), 385–404.

Schultze, C. L. *The Public Use of Private Interest*. Washington, D.C.: Brookings Institution, 1977.

Schwartz, B. The Effect in Philadelphia of Pennsylvania's Increased Penalties for Rape and Attempted Rape. *Journal of Criminal Law, Criminology and Police Science*, 1968, *59*, 509–515.

Schwartz, R. D., and Orleans, S. On Legal Sanctions. *University of Chicago Law Review*, 1967, *34*, 274–300.

Sethi, S. P. Liability without Fault? The Corporate Executive as an Unwitting Criminal. *Employee Relations Law Journal*, 1978, *4*, 185–219.

Skolnick, J. H. *Justice Without Trial: Law Enforcement in Democratic Society*. New York: Wiley, 1966.

Skolnick, J. H. Coercion to Virtue: Enforcement of Morals. *Southern California Law Review*, 1968, *41*, 588–641.

Skolnick, J. H. *Casino Gambling: The Erosion of Control*. Paper presented at the annual meeting of the Law and Society Association, San Francisco, May 11, 1979.

Smith, [N.I.]. *Ethical and Liability Problems in the Practice of Environmental Law*. Mimeograph, 1981.

Solomon, L. D., and Nowak, N. S. Managerial Restructuring Prospects for a New Regulatory Tool. *Notre Dame Lawyer*, 1980, *56*, 120–140.

Sproull, L. S. Response to Regulation: An Organizational Process Framework. *Administration and Society*, 1981, *12*, 447–470.

Staats, E. B. SMR Forum: Improving Industry–Government Cooperation in Policy Making. *Sloan Management Review*, 1980, *21*, 61–65.

Stafford, H. A. *The Effects of Environmental Regulations on Industrial Location*. Mimeograph, 1983.

Stewart, J., Anderson, E., and Taylor, Z. Presidential and Congressional Support for "Independent" Regulatory Commissions: Implications of the Budgetary Process. *Western Political Quarterly*, 1982, *35*, 318–326.

Stewart, R. B., and Krier, J. *Environmental Law and Policy: Readings, Materials and Notes*. Indianapolis: Bobbs-Merrill, 1978.

Stewart, R. B., and Sunstein, C. R. Public Programs and Private Rights. *Harvard Law Review*, 1982, *95*, 1193–1322.

Stigler, G. J. The Optimum Enforcement of Laws. *Journal of Political Economy*, 1970, *78*, 526–536.

Stigler, G. J. The Theory of Economic Regulation. *Bell Journal of Economics and Management Science*, 1971, *2*, 3–21.

Stone, C. D. *Where the Law Ends: The Social Control of Corporate Behavior*. New York: Harper & Row, 1975.

Stone, C. D. Social Control of Corporate Behavior. In M. D. Ermann and R. J. Lundman, Eds., *Corporate and Governmental Deviance: Problems of Organizational Behavior in Contemporary Society*. New York: Oxford University Press, 1978.

Stone, C. D. The Place of Enterprise Liability in the Control of Corporate Conduct. *Yale Law Journal*, 1980, *90*, 1–77.

Stone, C. D. Large Corporations and the Law at the Pass: Toward a General Theory of Compliance Strategy. *Wisconsin Law Review*, 1981, *5*, 861–890.

Stoney, D. J. The Economics of Environmental Law Enforcement, or Has the Prosecution of Pollutors Led to Cleaner Rivers in England and Wales? *Environment and Planning, A*, 1979, *11*, 897–918.

Stover, R. V., and Brown, D. W. Understanding Compliance and Noncompliance with Law: The Contributions of Utility Theory. *Social Science Quarterly*, 1975, *56*, 363–375.

Stover, R. V., and Brown, D. W. Reducing Rule Violations by Police, Judges and Corrections Officials. In S. S. Nagel, Ed., *Modeling the Criminal Justice System*. Beverly Hills, Cal.: Sage, 1977.

Sullivan, T. J. *Resolving Development Disputes Through Negotiations*. New York: Plenum Press, 1984.

Sumner, W. G. *Folkways*. New York: Arno Press, 1907.

Susskind, L., and Rubin, J. Negotiation–Behavioral Perspectives—Introduction. *American Behavioral Scientist*, 1983, *27*, 133–134.

Susskind, L., and Weinstein, A. How to Resolve Environmental Disputes Out of Court. *Technology Review*, 1982, *85*, 38.

Susskind, L., and Weinstein, A. Towards a Theory of Environmental Dispute Resolution. *Land Use and Environmental Law Review*, 1983, *14*, 433–479.

Susskind, L., Bacow, L., and Wheeler, M. *Resolving Environmental Regulatory Disputes.* Cambridge: Schenkman Publishing, 1983.

Swartzman, D., Liroff, R. A., and Croke, K. G., Eds. *Cost Benefit Analysis and Environmental Regulation: Politics, Ethics and Methods.* Washington, D.C.: Conservation Foundation Press, 1982.

Swigert, V., and Farrell, R. Corporate Homicide: Definitional Processes in the Creation of Deviance. *Law & Society Review,* 1980–1981, *15,* 161–182.

Tapp, J. L. Psychology and the Law: An Overture. *Annual Review of Psychology,* 1976, *27,* 359–404.

Tapp, J. L., and Kohlberg, L. Developing Senses of Law and Legal Justice. *Journal of Social Issues,* 1971, *27,* 65–91.

Tapp, J. L., and Levine, F. J. Persuasion to Virtue: A Preliminary Statement. In S. Krislov, K. O. Boyum, H. N. Clark, R. C. Schaefer, and S. O. White, Eds., *Compliance and the Law: A Multidisciplinary Approach.* Beverly Hills, Cal.: Sage, 1972.

Teubner, G. *After Legal Instrumentalism? Strategic Models of Post-Regulatory Law.* Paper presented at the annual Law and Society Association meeting, Boston, June, 1984.

Thompson, F. J. Deregulation by the Bureaucracy: OSHA and the Augean Quest for Error Correction. *Public Administration Review,* 1982, *42,* 202–212.

Thompson, G. P. Environmental Decision-Making: Agencies v. Courts. *Natural Resources Lawyer,* 1974, *7,* 367–371.

Tittle, C. R. Crime Rates and Legal Sanctions. *Social Problems,* 1980, *16,* 409–423.

Tittle, C. R., and Logan, C. H. Sanctions and Deviance: Evidence and Remaining Questions. *Law and Society Review,* 1973, *7,* 371–392.

Truitt, T. H., Berz, D. R., Weinberg, D. B., Molloy, J. B., Price, G. L., and Truitt, L. S. *Environmental Audit Handbook: Basic Principles of Environmental Compliance Auditing.* New York: Executive Enterprises, 1981.

Tucker, J. H. Corporate Compliance and Compensation Problems in Environmental Protection: Implications of United States versus Reserve Mining Company. *Natural Resource Lawyer,* 1977, *10,* 55–590.

Tunderman, D. W. Constitutional Aspects of Economic Law Enforcement. *Harvard Environmental Law Review,* 1980, *4,* 41–57.

Vaughan, D. Toward Understanding Unlawful Organizational Behavior. *Michigan Law Review,* 1982, *80,* 1377–1402.

Verkuil, P. R. Jawboning Administrative Agencies: Ex Parte Contacts by the White House. *Columbia Law Review,* 1980, *80,* 943–989.

Viscusi, W. K., and Zeckhauser, R. J. Optimal Standards with Incomplete Enforcement. *Public Policy,* 1979, *27,* 437–456.

Wald, P. Regulatory Negotiation: Its Place in the Emerging Field of Conflict Resolution. In ALI-ABA Course of Study Materials. *Environmental Law.* Philadelphia: American Law Institute, 1985.

Waldo, G. P., and Chiricos, T. G. Perceived Penal Sanctions and Self-Reported Criminality: A Neglected Approach to Deterrence Research. *Social Problems,* 1972, *19,* 522–540.

Wasby, S. L. The Supreme Court's Impact: Some Problems of Conceptualization and Measurement. In S. Krislov, K. O. Boyum, H. N. Clark, R. C. Schaefer, and S. O. White, Eds., *Compliance and the Law: A Multidisciplinary Approach.* Beverly Hills, Cal.: Sage 1972.

Wasby, S. L. *Small Town Police and the Supreme Court: Hearing the Word.* Lexington: Lexington Books, 1976.

Watson, J. L., and Danielson, L. J. Environmental Mediation. *Natural Resources Lawyer,* 1983, *15,* 687–723.

Weiner, N., and Mahoney, T. A. A Model of Corporate Performance as a Function of Environmental, Organizational and Leadership Influences. *Academy of Management Journal,* 1981, *24,* 452–470.

Werther, W. B. Government Control v. Corporate Ingenuity. *Labor Law Journal,* 1975, *26,* 360–367.

White, L. J. *Reforming Regulation: Processes and Problems.* Englewood Cliffs, N.J.: Prentice-Hall, 1981.

White, S. E., Dittrich, J. E., and Lang, J. R. The Effects of Group Decision-Making Processes and Problem-Situation Complexity on Implementation Attempts. *Administrative Science Quarterly,* 1980, *25,* 428–440.

Wichelman, A. F. Administrative Agency Implementation of the National Environmental Policy Act of 1969: A Conceptual Framework for Explaining Differential Response. *National Resource Journal,* 1976, *16,* 263–300.

Wilensky, H. L. *Organizational Intelligence.* New York: Basic Books, 1967.

Williams, D. L. Benefit–Cost Analysis of Natural Resources Decision-Making: An Economic and Legal Overview. *Natural Resources Lawyer,* 1979, *11,* 761–796.

Williams, K. R., and Gibbs, J. P. Deterrence and Knowledge of Statutory Penalties. *Sociological Quarterly,* 1981, *22,* 591–606.

Wilson, J. Q., Ed. *The Politics of Regulation.* New York: Basic Books, 1980.

Wilson, J. Q. Thinking About Crime. *The Atlantic,* 1983, *252,* 72–88.

Wilson, J. Q., and Rachal, R. Can the Government Regulate Itself? In M. D. Ermann and R. J. Lundman, Eds., *Corporate and Governmental Deviance: Problems of Organizational Behavior in Contemporary Society.* New York: Oxford University Press, 1978.

Wirt, F. M. *Politics of Southern Equality: Law and Social Change in a Mississippi County.* Chicago: Aldine Publishing, 1971.

Wolozin, H. The Economics of Air Pollution: Central Problems. *Law and Contemporary Problems,* 1968, *33,* 227–238.

Wright, J. P. *On a Clean Day You Can See General Motors: John Z. DeLorean's Look Inside the Automotive Giant.* Grosse Pointe: Wright Enterprises, 1979.

Zald, M. N., and Jacobs, D. Compliance/Incentive Classifications of Organizations: Underlying Dimensions. *Administration and Society,* 1978, *9,* 403–424.

Zeisel, H. *The Limits of Law Enforcement.* Chicago: University of Chicago Press, 1982.

Zimring, F., and Hawkins, G. Deterrence and Marginal Groups. *Journal of Research in Crime and Delinquency,* 1968, *5,* 100–114.

INDEX

Environmental law and American
business.